*Dedication*

*To my dearest parents, parents in-law, sweet sister, little brother*

*and*

*To my lovely wife*

# Practical Guide to Hot-Melt Extrusion: Continuous Manufacturing and Scale-up

Editor:
Mohammed Maniruzzaman

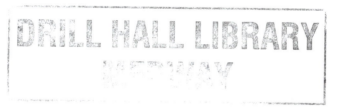

A Smithers Group Company

Shawbury, Shrewsbury, Shropshire, SY4 4NR, United Kingdom
Telephone: +44 (0)1939 250383  Fax: +44 (0)1939 251118
http://www.polymer-books.com

First Published in 2015 by

## Smithers Rapra Technology Ltd

Shawbury, Shrewsbury, Shropshire, SY4 4NR, UK

A catalogue record for this book is available from the British Library.

Every effort has been made to contact copyright holders of any material reproduced
within the text and the authors and publishers apologise if
any have been overlooked.

ISBN: 978-1-91024-265-0 (hardback)
978-1-91024-211-7 (softback)
978-1-91024-212-4 (ebook)

Typeset by Integra Software Services Pvt. Ltd.
Printed and bound by Lightning Source Inc.

# Preface

*Practical Guide to Hot-Melt Extrusion: Continuous Manufacturing and Scale-up* covers the main process technology (vertical over horizontal extrusion systems), operation principles and theoretical background of hot-melt extrusion (HME) techniques. It then continues by focusing on various novel applications (e.g., solubility enhancement, taste masking, controlled-release/sustained-release drug delivery). As per the recent trend of regulatory authority, taste masking has been an emerging topic and to date is considered a real challenge for the design and optimisation of drug products. The book then includes some recent and novel HME applications, case studies such as continuous co-crystallisations, use of novel excipients for continuous HME application and polymorphic transformation monitoring during the HME processing with respect to long-term stability and product performance. Scale-up considerations, case studies and regulatory issues during continuous HME processing are also included in this book.

Major topics covered include: the principles and process technology of HME in single-screw and twin-screw extrusion techniques; scale-up case studies; taste masking analysis using HME (*in vivo* and *in vitro*); solubility enhancements of poorly soluble active pharmaceutical ingredients using the HME technique as a means of continuous co-crystallisation; polymorphic transformation of crystalline drugs; devices used to process by HME; as well as providing an understanding into the regulatory perspectives relating to HME based products and their production methods. Various aspects of continuous manufacturing and the use of novel excipients for scale-up methodologies are discussed at the end of the book.

*Practical Guide to Hot-Melt Extrusion: Continuous Manufacturing and Scale-up* would be an essential multidisciplinary guide to the emerging pharmaceutical uses and applications for researchers in both academia and across industries working in the areas of drug formulation and delivery, polymers and materials sciences (pharmaceutical engineering and processing) for, (but not limited to) the following reasons:

- Emerging HME processes and applications for multiple drug delivery.

- Solid-state engineering, solvent free co-crystals manufacturing and solubility enhancement, controlled-release taste masking and sustained-release case studies from a continuous manufacturing viewpoint. Scale-up case study and a consideration of the issues relating to the regulatory guidelines for continuous manufacturing involving emerging HME techniques.

Dr. Mohammed Maniruzzaman

University of Greenwich, UK, 2015

# Contents

Preface........................................................................................................... iii

Contents........................................................................................................ vii

Contributors................................................................................................. xv

1   Introduction to Hot-Melt Extrusion, Continuous Manufacturing:
    Scale-up *via* Hot-Melt Extrusion ...................................................... 1

        1       Background................................................................................. 1

        1.1     Introduction............................................................................... 1

                1.1.1   Hot-Melt Extrusion Process Technology ................. 2

                1.1.2   Equipment .................................................................. 3

                1.1.3   Benefits and Drawbacks to Hot-Melt Extrusion ...... 4

                1.1.4   Application of Hot-Melt Extrusion ......................... 5

                1.1.5   Materials used in Hot-Melt Extrusion Processes ..... 6

                1.1.6   Development of Sustained-release Formulations...... 7

                1.1.7   Marketed Products..................................................... 7

        1.2     Continuous Manufacturing...................................................... 8

        1.3     Scale-up *via* Hot-Melt Process............................................. 10

                1.3.1   Case Study: Scale-up of a Hot-Melt Extrusion
                        Process using Soluplus® as Carrier........................ 11

        1.4     Conclusions ........................................................................... 14

        References ........................................................................................... 15

2   Hot-Melt Extrusion as Continuous Manufacturing Technique,
    Current Trends and Future Perspectives ........................................ 19

        2.1     Hot-Melt Extrusion as Continuous Manufacturing
                Technique................................................................................ 19

                2.1.1 Hot-Melt Extrusion Equipment ................................ 21

                2.1.2 Single-screw Extruder ................................................ 24

2.1.3      Twin-screw Extruders.................................................. 24

2.1.4      Role of Glass Transition Temperature ................... 26

2.1.5      Role of Mechanical Parameters ............................. 26

2.1.6      Use of Plasticisers or Surfactants ......................... 26

2.1.7      Challenges and Opportunities ............................... 26

2.2      Background Information on Batch Processing...................... 27

2.2.1      Past and Future Trends of Continuous
Manufacturing ...................................................... 28

2.2.2      Conventional Batch Process Design....................... 29

2.2.3      Production Parameters ......................................... 29

2.2.4      Production Parameters in Batch Processing .......... 29

2.2.4.1      Active Pharmaceutical Ingredient(s)
Manufacturing Site ........................... 29

2.2.4.2      Formulation Development................... 29

2.2.4.3      Industrial Scale-up ............................. 30

2.2.4.4      Regulatory Guidelines........................ 30

2.2.5      Validation of Batch Processing ............................. 30

2.2.6      In-line Continuous Monitoring Techniques .......... 30

2.3      Assimilation of Complete Quality-by-Design Models............ 35

2.3.1      Chief Modification in Procedural Expertise
for Continuous Manufacturing ............................. 35

2.3.2      Process Parameters for Preformulation for
Hot-Melt Extrusion............................................... 36

2.3.3      Implementation of a New Product
Development Process.............................................. 36

2.3.3.1      Modelling Simulation of
Hot-Melt Extrusion Process ............... 36

2.3.4      Flow Chart Model for Continuous
Hot-Melt Extrusion based Pharmaceutical
Manufacturing Process ......................................... 39

2.3.5      Foremost Alteration in Structural Provisions for
Quality and Technical Operations......................... 41

2.3.6      Potential Benefits of Continuous
Hot-Melt Extrusion Manufacturing ..................... 44

2.3.7      Regulatory Issues Related to
Hot-Melt Extrusion............................................... 44

|  |  | 2.3.8 | Process Analytical Technology Framework | 45 |
|  | 2.4 | | Conclusion | 46 |
|  | | | References | 47 |

3 Co-extrusion as a Novel Approach in Continuous Manufacturing Compliance ... 53

|  | 3 | | Background | 53 |
|  | 3.1 | | Introduction | 53 |
|  | 3.2 | | Applications of Co-extrusion *via* Hot-Melt Extrusion in Continuous Manufacturing | 54 |
|  | | 3.2.1 | Problem Encountered During Co-extrusion | 55 |
|  | | 3.2.2 | Variation of the Die Temperature | 55 |
|  | | 3.2.3 | Swelling of the Die | 55 |
|  | | 3.2.4 | Viscosity Matching | 56 |
|  | | 3.2.5 | Adhesion | 56 |
|  | | 3.2.6 | Interdiffusion | 56 |
|  | | 3.2.7 | Delamination | 57 |
|  | 3.3 | | Pharmaceutical Applications | 57 |
|  | | 3.3.1 | Pharmaceutical Significance of the Co-extrusion Technique | 57 |
|  | 3.4 | | Case Study | 58 |
|  | 3.5 | | Conclusions | 59 |
|  | | | References | 59 |

4 Solid-state Engineering of Drugs using Melt Extrusion in Continuous Process ... 63

|  | 4 | | Introduction | 63 |
|  | 4.1 | | Dissolution Enhancement | 63 |
|  | | 4.1.1 | Use of Novel Inorganic Excipients | 63 |
|  | | 4.1.2 | Use of Polymers to Enhance Dissolutions of Poorly Water-soluble Active Pharmaceutical Ingredients | 67 |
|  | | 4.1.3 | Continuous Co-crystallisation Engineering | 68 |
|  | | 4.1.4 | Polymorphic Transformations *via* Hot-Melt Extrusion | 69 |

           4.1.4.1    Polymorphic Transformation of Carbamazepine ................................... 70

           4.1.4.2    Polymorphic Transformation of Artemisinin ......................................... 70

     References ...................................................................... 71

5    Continuous Co-crystallisation of Poorly Soluble Active Pharmaceutical Ingredients to Enhance Dissolution ................................. 75

     5.1    Introduction ................................................................. 75

     5.2    Mechanism of Continuous Co-crystallisation by Twin-screw Extrusion ......................................................... 79

         5.2.1    Eutectic-mediated Co-crystallisation ..................... 80

           5.2.1.1    Ibuprofen-Nicotinamide Case ............. 80

           5.2.1.2    Carbamazepine-Saccharin Case........... 81

         5.2.2    Solvent-assisted Co-crystallisation........................ 85

         5.2.3    Amorphous Phase-mediated Co-crystallisation...... 87

           5.2.3.1    AMG 517-Sorbic Acid Case............... 87

     5.3    Critical Parameters Influencing Continuous Co-crystallisation by Twin-screw Extrusion .......................... 89

         5.3.1    Processing Temperature......................................... 89

         5.3.2    Temperature-dependent Case ................................ 89

         5.3.3    Temperature-independent Case ............................. 89

         5.3.4    Screw Design ........................................................ 90

         5.3.5    Screw Speed ......................................................... 90

         5.3.6    Feed-rate .............................................................. 91

     5.4    Case Study .................................................................. 91

         5.4.1    Matrix-assisted Co-crystallisation of Carbamazepine-Nicotinamide System by Twin-screw Extrusion................................................ 91

         5.4.2    Avoiding Thermal Degradation of Carbamazepine through *In Situ* Co-crystallising with Nicotinamide ....................... 92

     5.5    Conclusions ................................................................. 93

     References ...................................................................... 93

6 Taste Masking of Bitter Active Pharmaceutical Ingredients for the Development of Paediatric Medicines *via* Continuous Hot-Melt Extrusion Processing ...... 97

6 Introduction ...... 97

6.1 Hot-Melt Extrusion as an Active Taste Masking Technique ...... 97

6.1.1 Taste Masking *via* Continuous Hot-Melt Extrusion Process ...... 97

6.1.2 Polymers as Suitable Carriers for Taste Masking ...... 99

6.1.3 Use of Lipids as Carriers for Taste Masking in Hot-Melt Extrusion ...... 104

6.2 Continuous Manufacturing of Oral Films *via* Hot-Melt Extrusion ...... 105

6.2.1 Case Study: Taste Masking of Bitter Active Pharmaceutical Ingredients *via* Continuous Hot-Melt Extrusion Processing ...... 106

6.3 Materials ...... 107

6.3.1 *In Vivo* Taste Masking Evaluation ...... 107

6.3.2 *In Vitro* Taste Masking Evaluation: Astree e-tongue ...... 107

6.3.3 Sample Preparation for Astree e-tongue ...... 108

6.4 Results and Discussion ...... 108

6.4.1 *In Vivo/In Vitro* Taste Evaluations ...... 111

6.5 Conclusions ...... 115

References ...... 115

7 Continuous Manufacturing of Pharmaceutical Products *via* Melt Extrusion: A Case Study ...... 121

7 Introduction ...... 121

Case Study ...... 122

7.1 Materials and Method ...... 122

7.1.1 Materials ...... 122

7.1.2 Preparation of Formulation Blends and Continuous Hot-Melt Extrusion Processing ...... 122

7.1.3 Scanning Electron Microscopy/Energy Dispersive X-ray Analysis ...... 123

| | | 7.1.4 | Thermal Analysis | 123 |
| | | 7.1.5 | Powder X-ray Diffraction | 124 |
| | | 7.1.6 | *In Vitro* Drug Release Studies | 124 |
| | | 7.1.7 | Analysis of Drug Release Mechanism | 124 |
| | 7.2 | | Results and Discussion | 126 |
| | | 7.2.1 | Continuous Manufacturing of Pellets *via* Hot-Melt Extrusion | 127 |
| | | 7.2.2 | Advanced Surface Analysis | 129 |
| | | 7.2.4 | Thermal Analysis | 130 |
| | | 7.2.5 | Amorphicity Analysis | 131 |
| | | 7.2.6 | *In Vitro* Dissolution Studies | 132 |
| | | 7.2.7 | Analysis of Release Mechanism | 133 |
| | 7.3 | | Conclusions | 135 |
| | | | References | 136 |
| 8 | Novel Pharmaceutical Formulations Using Hot-Melt Extrusion Processing as a Continuous Manufacturing Technique | | | 139 |
| | 8 | | Introduction | 139 |
| | 8.1 | | Materials and Methods | 140 |
| | | 8.1.1 | Materials | 140 |
| | | 8.1.2 | Hot-Melt Extrusion Processing | 140 |
| | | 8.1.3 | Differential Scanning Calorimetry Analysis | 140 |
| | | 8.1.4 | Hot Stage Microscopy Analysis | 140 |
| | | 8.1.5 | X-ray Powder Diffraction | 141 |
| | | 8.1.6 | *In Vitro* Drug Release Studies | 141 |
| | | 8.1.7 | High-performance Liquid Chromatography Analysis | 141 |
| | 8.2 | | Results and Discussion | 142 |
| | | 8.2.1 | Hot-Melt Extrusion Processing | 142 |
| | | 8.2.2 | Thermal Analysis | 142 |
| | | 8.2.3 | X-ray Powder Diffraction Analysis | 144 |
| | | 8.2.4 | *In Vitro* Dissolution Studies | 145 |
| | 8.3 | | Conclusions | 146 |
| | | | References | 147 |

9   Continuous Polymorphic Transformations Study *via* Hot-Melt Extrusion Process ................................................................ 149

    9    Background ............................................................................ 149

        9.1    Polymorphism ........................................................... 149

        9.2    Case Study: Polymorphic Transformation of Paracetamol .............................................................. 150

               9.2.1    Experimental Methods .................................. 151

               9.2.2    Theoretical Calculation ................................ 151

               9.2.3    Continuous Hot-Melt Extrusion Process and In-line Monitoring ...................................... 152

               9.2.4    Thermal Analysis ......................................... 153

        9.3    Results and Discussion .............................................. 154

               9.3.1    Continuous Extrusion Process and Theoretical Consideration ............................. 154

               9.3.2    Physicochemical Characterisation of the Polymorphic Transformation during Hot-Melt Extrusion ...................................... 156

               9.3.3    In-line Near-infrared Spectroscopy Monitoring .................................................. 163

        9.4    Conclusions .............................................................. 165

    References .................................................................................. 165

10   From Pharma Adapted Extrusion Technology to Brand New Pharma Fitted Extrusion Design: The Concept of Micro-scale Vertical Extrusion and its' Impact in Terms of Scale-up Potential ............ 169

    10   Introduction ......................................................................... 169

        10.1   The Advantages of Hot-Melt Extrusion ............................ 170

              10.1.1  Examples of Problems Solved Thanks to Hot-Melt Extrusion ...................................... 170

        10.2   From Micro-scale to Industrial-scale ................................ 171

              10.2.1  Viscosity of the Three Samples at Different Temperatures ............................................... 172

              10.2.2  Viscosity at Various Temperatures .................. 173

        10.3   Hot-Melt Extrusion as a Standard Technology for the Pharmaceutical Industry ............................................. 174

        10.4   Innovations ............................................................. 175

| | | |
|---|---|---|
| 10.5 | Conclusions | 179 |
| | References | 179 |
| 11 | Continuous Manufacturing *via* Hot-Melt Extrusion and Scale-up: Regulatory Aspects | 181 |
| 11 | Introduction | 181 |
| 11.1 | Continuous Manufacturing and Hot-Melt Extrusion Processing Technology | 181 |
| 11.2 | Aspects of the Controls/Parameters in Hot-Melt Extrusion Processing | 184 |
| 11.3 | Continuous Manufacturing *via* Hot-Melt Extrusion | 185 |
| 11.4 | Continuous Manufacturing Process over Batch Process | 187 |
| 11.5 | Scale-up Methodologies | 188 |
| 11.6 | Regulatory Aspects | 191 |
| 11.7 | Summary | 193 |
| | References | 193 |
| Abbreviations | | 197 |
| Index | | 201 |

# Contributors

**Ritesh Fule**

Department of Pharmaceutical Science and Technology, Institute of Chemical Technology, Mumbai, 400019, India


**Ming Lu**

Department of Pharmaceutics, School of Pharmaceutical Sciences, Sun Yat-sen University, Guangzhou, China


**Hans Maier**

Rondol Industrie SAS, 8, Place de l'Hôpital, 67000, Strasbourg, France


**Mohammed Maniruzzaman**

University of Greenwich, Faculty of Engineering and Science, Department of Pharmaceutical Sciences, Chatham Maritime, Chatham, Kent, ME4 4TB, UK


**Victoire de Margerie**

Rondol Industrie SAS, 8, Place de l'Hôpital, 67000, Strasbourg, France

# 1 Introduction to Hot-Melt Extrusion, Continuous Manufacturing: Scale-up *via* Hot-Melt Extrusion

Mohammed Maniruzzaman

## 1 Background

Extrusion is a word, originating from the Latin *extrudere*, which means to press out or to drive out [1]. By definition, extrusion is a technique to drive out a new complex material known as an extrudate, by forcing a mixture through a die under controlled conditions such as temperature, mixing speed, feed-rate and pressure [2]. Application of this technique by the industry started back in the 1930s. Until then, (mainly from the mid-1970s and onwards), it was considered one of the most practical techniques used for the processing of food, plastics and rubber by industries around the world. Estimated statistics shows that more than half of the products of these three categories are manufactured by this process. Nowadays, this technique has been used as a production method in the pharmaceutical industry [3].

## 1.1 Introduction

The extrusion process is now used to produce a variety of dosage forms and formulations such as granules, pellets, tablets, suppositories, implants, stents, transdermal systems and ophthalmic inserts. For this reason, hot-melt extrusion (HME) is an excellent alternative technique for use in place of conventional techniques such as solvent evaporation, freeze-drying, and spray drying and so on. This continuous manufacturing technique can increase the solubility of poorly-soluble pharmaceutical actives and can enhance the bioavailability of the respective drug by forming a solid dispersion or solid solution [4]. Early researches of HME processing have described the preparation of matrix mini-tablets which was followed by further investigations into the properties of sustained-release (SR) mini-matrices tablets [5, 6]. Extruded mini-tablets showed minimised risk of dose dumping and reduced inter- and intra-subject variability. Rather than a proven manufacturing process of pharmaceutical dosage forms, this technology also meets the rules and regulations of the US Food and Drug Administration's (FDA) process analytical technology (PAT) scheme for designing, analysing as well as controlling the manufacturing process *via* quality control measurements during active extrusion process [7]. Nowadays, scientists are also making chrono-pharmaceutical dosage forms which are designed to release the drug at a specific time to improve the patient's compliances and drug's therapeutic efficacy. A chrono-pharmaceutical dosage form is also known as pulsatile release or time release dosage forms. The HME technique can be applied to manufacture these types of SR formulations [8]. For drugs which are processed by conventional techniques, it is difficult to control their release rate and thus variation occurs in the oral bioavailability [9]. However, multiple unit dosage forms can overcome this problem to a certain extent, using HME to develop a new dosage form to overcome these problems.

### 1.1.1 Hot-Melt Extrusion Process Technology

The use of HME technology (invented by J. Brama in the early 1930s) was first introduced to the pharmaceutical industry by Dolker and co-workers, followed by Breitenbach [10]. The first manufacturing of SR dosage form of various types of drugs were developed by Follonier and his co-workers [11]. In this technique, drugs and polymers are melted together to form a homogeneous mixture and transformed into different types of finished products [12]. The whole process can be briefly explained by the following steps as shown in **Figure 1.1** [13]:

- Material feeding through hopper Mixing, grinding, reducing the particle size, venting and kneading

- Flow through the die

- Extrusion from the die and further downstream processing

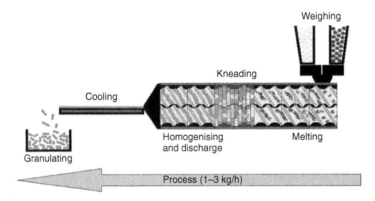

**Figure 1.1** Schematic diagram of a hot-melt extruder. Reproduced with permission from M. Maniruzzaman, J.S. Boateng, M.J. Snowden and D. Douroumis, *International Scholarly Research Notices: Pharmaceutics*, 2012, Article ID:436763. ©2012, M. Maniruzzaman, J.S. Boateng, M.J. Snowden and D. Douroumis [13]

An extruder generally consists of one or two screws inside a heated cylindrical barrel. A hopper is attached at one side of the barrel to feed the material towards the screw and whilst on the other side, a die is attached to transform the softened material into the desired size and shape [14].

## 1.1.2 Equipment

A basic single-screw extruder contains one rotating screw positioned inside a stationary barrel. The temperature of the stationary barrel can be controlled by external heating equipment. However, in an advanced twin-screw extruder, two co-rotating or counter-rotating screws aid the extrusion process [15]. In this type of extruder, screws must be capable of rotating in a defined rpm by regulating a fixed torque and shear force. Apart from this screw, an extruder machine also contains a motor which acts as a drive unit, a stationary barrel, a die, a conveyor and cutting equipment [3]. Process parameters such as screw speed, temperature and pressure are controlled by a central controlling unit having a monitoring device [16]. Screws have a specific geometry according to the length and diameter (L/D) ratios (**Figure 1.2**). The L/D of the screw of a single-screw or a twin-screw extruder can typically range from 20 to 40:1 in mm. In a commercial type of extruder, the diameter of a screw may be between 50—66 mm [17]. A basic design of an extruder machine consists of three discrete zones: feed zone, compression zone and a metering zone [14]. Normally the processing pressure in the feed zone stays very low but as the materials flow towards the mixing zone, pressure increases greatly. To compensate this matter, the depth and pitch stays much deeper compared to the other zones of the screws. This increase in pressure homogenises the material.

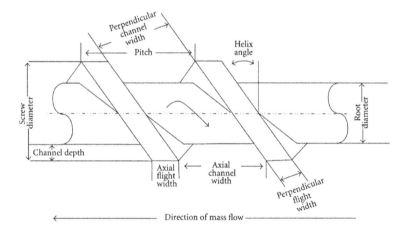

Figure 1.2 Screw geometry. Reproduced with permission from M. Maniruzzaman, J.S. Boateng, M.J. Snowden and D. Douroumis, *International Scholarly Research Notices: Pharmaceutics*, 2012, Article ID:436763. ©2012, M. Maniruzzaman, J.S. Boateng, M.J. Snowden and D. Douroumis [13]

The screws' pitch and flight depth play an important role for this homogenisation process [17]. The metering zone ensures the uniform thickness, shape and size of the extrudates. Constant screw geometry can raise high-pressure, providing a uniform distribution of material in the extrudates.

There is some additional auxiliary equipment are also installed in an extruder machine for cooling, cutting and collection of the pellets. In order to analyse the samples, various analytical techniques such as near-infrared (NIR), Raman, ultrasound, differential scanning calorimetry system, and thermocouples can be used to monitor the process in real time.

### 1.1.3 Benefits and Drawbacks to Hot-Melt Extrusion

The HME process has certain benefits and drawbacks which are given below [1].

Benefits:

- The HME process enhances the bioavailability of poorly soluble components.

- Processing of materials can be done in the absence of solvents and water.

- HME is an economical process with reduced production time, less processing steps and a continuous operation.

- Sustained-, modified- and targeted-release formulations can be made using this novel technique.

- Better content uniformity can be obtained from the HME processing method.

- Compressibility of active ingredients can be readily increased and the entire process is simple, continuous and efficient.

- Uniform solid dispersions can be achieved by this method.

- HME increases the stability of the finished product at varying pH and moisture levels.

- The number of unit operation steps greatly decreases.

- HME can be used to manufacture a wide range of pharmaceutical dosage forms.

Drawbacks:

- HME is a non-ambient processing technique.

- Flow properties of the polymers are essential for processing.

- The number of polymers suitable for processing is limited.

- HME requires a high energy input.

- This technique cannot be used for heat sensitive materials.

- Polymers having lower melting points can cause softening of the finished dosage form and agglomerates can be formed during storage.

- Polymers having higher melting point can cause degradation of the heat sensitive material.

### 1.1.4 Application of Hot-Melt Extrusion

Extrusion technology is one of the most robust techniques used as a fabrication process. Insulated wires, plastic pipes, cables, rubber sheets and polystyrene tiles can be made by using this technology [18]. It is extensively used in the food industry for the production of spaghetti and pasta [1]. In the production of pelletised feed, implants and injection moulding, this process has been used for ages [19].

HME technology has received great attention in the pharmaceutical industry and academia due to its easy processing steps and efficiency. This technology enhances the quality of the manufactured product [20]. Pharmaceutical active ingredients can be dispersed at a molecular level by forming solid dispersions using this technology [2]. HME can be used to enhance dissolution rate and bioavailability of poorly soluble drugs. Increasing the solubility of poorly water-soluble drugs is a real challenge in the formulation development process [21]. Today, the discovery of high molecular weight molecules is achieved through high throughput screening which shows up very low solubility and therefore poor bioavailability [22]. Control or modification of the release, masking the taste of the drug, and the production of thin films are the various types of application of HME technology [23]. HME has countered solubility issues by preparing molecular dispersions of active pharmaceutical ingredients (API) into different hydrophilic polymer matrices [21, 22].

### *1.1.5 Materials used in Hot-Melt Extrusion Processes*

The API and excipients should be safe, pure and non-toxic for the development of dosage delivery forms through HME. Most of the compounds used in the HME formulations, are taken from the conventional formulation designs. The basic property of the excipients is thermal stability in addition to acceptable physical and chemical stability. HME formulations are complex mixes of pharmaceutical active ingredients. These can be categorised as matrix carriers, release-modifying agents, bulking agents and additives [24].

HME is a comparatively new technique compared to traditional formulation development. It offers benefits over conventional techniques of manufacturing. HME is an anhydrous process, so there is no chance of having a hydrolytic degradation of the drug product. It increases the compactability of poorly compactable material. As a robust mixing happens throughout the process, sometimes drug particles dissolve into the excipient material or remain undissolved as a solid solution in the final dosage form. Sometimes the molten drug changes the overall property of the extrudates. For example, molten oxprenolol hydrochloride reduced the viscosity of the extrudate and exhibited poor handling property [25]. Polymer system selection is also a very big issue when designing a proper formulation. This system may be of polymeric material or low melting waxes [25, 26] including polyvinyl pyrrolidone (PVP) or its co-polymer such as PVP-vinyl acetate, poly(ethylene-*co*-vinyl acetate) (EVAC), various grades of polyethylene glycols (PEG), cellulose ethers and acrylates, various molecular weight of polyethylene oxides, polymethacrylate derivatives and poloxamers. Amongst the different classes of biodegradable polymers, the thermoplastic aliphatic polyesters such as polylactide, polyglycolide and copolymer of lactide and glycolide, poly(lactide-*co*-glycolide) have been used as carriers in the extrusion process [4]. Various types of meltable binders are also used to develop SR dosage forms i.e., PEG, Eudragit and so on. The use of polymeric carriers sometimes causes hardening of the extrudates. In this case, a plasticiser can be added to the formulation to improve the processing conditions during the manufacturing of the final product [27]. The use of a plasticiser increases the flexibility of the extrudates. Plasticisers also lower the glass transition temperature of the polymer and improved temperature conditions improve the stability profile of the active compound or polymer carrier [24, 28]. The shear forces can be compromised by using a plasticiser but thermo-chemical stability and volatility should also be taken into consideration before the selection of the right excipients. There are some other polymers which can stabilise the system including antioxidants, acid receptors or light absorbers. Antioxidants prevent the initiation of free radical chain reaction. Reducing agents such as ascorbic acid can offer such prevention to drugs polymers and other excipients. The most effective is to use a closed system to preserve the formulation. The rate of radical ion formation can also be minimised by the use of edetate disodium or citric acid.

### 1.1.6 Development of Sustained-release Formulations

Follonier and co-workers started the investigation of producing SR pellets in 1994 [29]. The study was carried out using ethyl cellulose (EC), cellulose acetate butyrate, polyethyl acrylate/methyl methacrylate/trimethyl ammonio ethyl methacrylate chloride (Eudragit RSPM) and EVAC. Triacetin and diethyl phthalate were used as plasticisers. The result was not so impressive that time with 65–80% of the drug being released over a 6-h time period. HME process and formulation design is also used to develop gastro-resistant tablets. An enteric coating can be produced at a molecular level by using acrylate polymers such as Eudragit L [30]. Further research on polyethylene oxide (PEO) found that the stability of the formulation was dependent on the storage conditions, processing temperature and the molecular weight of the polymer. Sometimes the degradation can be accelerated by using low molecular weight PEO. The drug release mechanism from EC matrix tablets depends on particle size, compaction force and extrusion temperature. So, it is clear that a number of variables can affect the HME formulation development process.

Now researchers have been focusing on the development of mini-matrices by solid dispersion of the drug molecules into the polymer matrices *via* HME technology [31]. Further analysis of these mini-matrices showed the properties of SR dosage forms, composed of EC, HPMC and Ibuprofen, as explained by De Brabander and co-workers (2000) [5, 6]. These mini-matrices have certain advantages including a minimised risk of dose dumping and reduced variability. Retarded-release pellets, by using vegetable calcium stearate as an excipient through HME technology, was developed by Roblegg and his co-workers [32]. HME can also be used to develop a controlled-release system based on Acryl-EZE® [33].

### 1.1.7 Marketed Products

From the early 1980s, HME-related patents have steadily increased [34]. Up to date, the US and Germany hold approximately 56% of all issued patents [4]. Renowned pharmaceutical industries such as Pharma Form (USA), SOLIQS (Germany) are using this drug delivery technique. Meltrex and Kaletra are two great examples of finished products made by this technology. Kaletra tablets showed better patient compliance in terms of reduced dosing frequency and improved stability compared to the previous soft gel formulation. On the other hand, a fast onset Ibuprofen formulation and a SR formulation of verapamil formulation were developed by SOLIQS. Through HME technology, SOLIQS introduced Meltrex® technology which is a calendering process to directly shape the extrudates into a tablet form.

## 1.2 Continuous Manufacturing

The term 'continuous manufacturing' refers to a process whereby material is simultaneously charged and discharged (e.g., petroleum refining). In batch manufacturing, all materials are generally charged before the start of processing and discharged at the end of processing which results in a process which is very lengthy. In batch manufacturing, the product is most frequently tested off-line after processing is complete, so there is a lack of opportunities for real time monitoring during the operational process. Extensive research and development have led continuous manufacturing to leave the trial phase and to get a place in commercial deployment. It is therefore encouraged to commence this method of continuous us manufacturing which can offer numerous benefits such as integrated processing with fewer steps – no manual handling, increased safety and shorter processing times In addition, smaller equipment and facilities, more flexible operation, lower capital costs and rapid development screening over many conditions, are considered important factors for adapting continuous manufacturing in pharmaceutical manufacturing. Continuous manufacturing also provides scope for extended on-line monitoring and product quality assurance in real-time (FDA) [35–37]. In the industrial production of drugs, active substances rarely have the physical requirements needed for processing and manufacturing the final products and therefore the potential of functional excipients have being revolutionising the pharmaceutical industries.

Many new drugs have poor physicochemical and biopharmaceutical properties which constrain bioavailability, processing and clinical performance. The reduction in the numbers of new molecules coming to market along with the expiry of patents is a major concern for the pharmaceutical industry. This often results in repositioning and reformulation of the drugs. There is a clear market need to develop methods to make drugs more soluble. It has been reported that about 40% of the drugs in the discovery pipeline are poorly soluble or insoluble thereby increasing the cost and time of drug development [38]. Similarly, large numbers of drugs have stability and compressibility issues, solutions which are currently being explored by crystal and particle engineering techniques. There are various potential methods to overcome issues such as solubility in drug delivery strategy. These include particle engineering to reduce the size of the drug particles and forming salts or solid solutions by dissolving the drug in a soluble polymer. These techniques are generally complex and are only suitable for some types of drugs. Another potential method of improving the solubility of certain drugs is to form a co-crystal of the drug with another pharmaceutically accepted material, such as a sugar. Recently, there has been a significant emphasis on the effort to make the drugs available on a more patient-specific basis where personalised medicines come to the forefront to tackle this issue. Suitable process optimisation for appropriately chosen drug delivery strategy can overcome this issue significantly. A suitable example can be use of three-dimensional printing as a method of producing personalised medicines where as and when needed dosage forms can be manufactured and provided to patients saving time and providing direct benefits to the public health. Such a step forward will involve a combined approach with advanced drug delivery and process engineering. There is a tremendous need to explore this opportunity to adopt a continuous mode of manufacturing that can cut the production cost down significantly.

Recently, it has been reported that HME has been successfully utilised in manufacturing pharmaceutical products for multiple drug delivery strategy [39–43]. HME can either be processed with single-screw or twin-screw instruments. Generally, single-screw extruder has single-screw with conveying properties built as close to the barrel as possible to produce sufficient shear. In contrast, twin-screw extruders can be adapted with various screw designs, configurations as per the requirement of formulations and the final dosage forms. Screw geometry of twin-screw (co-rotating or counter-rotating) produces the highest shear and offers excellent mixing capacity in the barrel. The main advantage of HME techniques includes the viability for continuous processing without the use of solvents and the fact that it is' economically friendly and easy to scale-up.

A new terminology which is being highly encouraged by the regulatory authority to have adopted is known as 'quality-by-design' (QbD), which is described in [44]. International Conference on Harmonisation (ICH) guidelines in various paragraphs such as ICH Q2 (R2), ICH Q9 and ICH 10 [44]. The main objectives of QbD approaches are to focus on science based design and development of formulation and manufacturing processes in order to ensure predefined product quality objects to develop design space [45]. To ensure the product quality and in-line measurement of critical product parameters, FDA has started an initiative known as PAT, which is normally used to control and understand the manufacturing process [46]. NIR spectroscopy is one of the most common techniques which is suitable for an increased number of PAT applications and has been used in several studies in the pharmaceutical and nutritional fields [47]. Dhumal and co-workers and Kelly and co-workers used NIR as a PAT tool to monitor co-crystal formation during a solvent free continuous crystallisation process [10, 48]. However, mostly these reported studies have primarily focused on the manufacture of drug materials with improved quality but none of them actually looked at the potential utilisation of these manufactured drugs into pharmaceutical formulations and products such as tablets or biodegradable implants. Still, there is an immense need for using continuous manufactured drugs in final products *via* one-step processing (e.g., HME). This will have a significant economic impact on pharmaceutical industries worldwide as well as on public health through the feasibility of otherwise unusable drugs.

Being adopted as a suitable process engineering platform, HME can be optimised for use as a continuous manufacturing technique for pharmaceutical formulations and products where screw speed, feeding rate, temperature profile and screw configurations can be taken as the critical processing parameters. These parameters impact on the manufactured formulations' solubility, dissolution rate, particle size and so on. These are considered as critical quality attributes. Having considered the critical quality attributes and the critical processing parameters, the QbD approach can be then implemented *via* the outlined set-up in HME to ensure final quality product and develop a design space.

## 1.3 Scale-up *via* Hot-Melt Process

Pilot scale manufacturing requires a high throughput rate of materials that cannot be met by using laboratory scale HME processing. Therefore, process scale-up is required to enable large scale manufacturing whilst maintaining the critical quality attributes of the product. A thorough investigation of the scale-up processes in a continuous manufacturing platform using HME can be optimised using a numbers of models and theories. In literature, various case studies have been reported using an adiabatic melt extrusion process derived from cubic law [49]. Similarly, a square law of scale-up can also be proposed, based on the heat transfer area [50]. However, establishing the balance on the geometric similarity and the melting temperature ($T_m$) may not be constant before and after scale-up. Later on, a more advanced adiabatic index in the scale-up process had been proposed and a theory developed which was aimed at keeping the $T_m$ constant [51, 52].

*Volumetric scale-up*: Volumetric scale-up of the continuous manufacturing of pharmaceutical products using the HME process will focus on maintaining the same degree of fill and consequently, the same residence time distribution (RTD). For geometrically similar extruders with different diameters, the targeted process throughput will be estimated using the **Equation 1.1** [53]:

$$Q_T = Q_M \times \frac{D_T^3}{D_M^3} \times \frac{N_T}{N_M} \tag{1.1}$$

where $Q_T$ and $Q_M$ are the targeted and initial process throughput, respectively; $D_T$ and $D_M$ are the screw diameters after and before scale-up, respectively; and $N_T$ and $N_M$ are the screw speed after and before scale-up, respectively. In a simplified situation where the volumetric scale-up is conducted using similar geometric extruders (e.g., an 11 mm or 16 mm twin-screw extruder from Thermo Fisher, Germany), the screw speed is to be kept constant and the process throughput is expected to obey cubic law.

*Scale-up via heat transfer*: Scale-up of the continuous manufacturing of pharmaceutical pharmaceuticals using HME process can be assessed by optimising the heat transfer of the barrel where the scale-up of the product manufacturing will

entirely be dependent on the heat transfer. In this method, the surface area for heat transfer will be taken as equivalent to the barrel surface area. Based on this, the process throughput will be scaled up while maintaining the heat transfer rate by following **Equation 1.2** [53]:

$$Q_T = Q_M \times \frac{D_T^2}{D_M^2} \times \frac{N_T}{N_M} \qquad (1.2)$$

where $Q_T$ and $Q_M$ are the targeted and initial process throughput, respectively; $D_T$ and $D_M$ are the targeted and initial screw diameter, respectively; and $N_T$ and $N_M$ are the screw speed after and before scaling up, respectively. From this, it can be found that the heat transfer scale-up follows the square law if the screw speed does not change.

*Power scale-up*: The specific torque and specific energy (SE) input is considered a critical parameter in the production of high-energy compositions including amorphous systems. Therefore, a successful scale-up in continuous manufacturing process *via* HME depends to a large extent on the steady constant level of SE input. Generally, the mechanical energy input in a HME process during the optimisation of scale-up methods, is calculated and determined by using an **Equation 1.3** as described below:

$$E = E_{max} \times \frac{N}{N_{max}} \times \frac{\tau}{\tau_{max}} \times Gearbox\,Rating \qquad (1.3)$$

where $E_{max}$, $N_{max}$ and $\tau_{max}$ of a large scale extruder is normally predetermined by the instrument design. The exact value of maximum torque and screw speed is measured directly. Therefore, the mechanical energy of large scale equipment can be calculated and used to determine the process throughput which maintains the SE input as constant. For extrusion systems with the same specific torque, the process throughput can be scaled up by following the cubic law while keeping the SE input unchanged if the screw speed is constant [53].

### 1.3.1 Case Study: Scale-up of a Hot-Melt Extrusion Process using Soluplus® as Carrier

A study, recently reported by Gryczke and co-workers (2013) investigated and provided a deeper understanding of the influence of process parameters on the RTD of the material within the extruder and the specific mechanical energy consumption (SMEC) during the extrusion process when Soluplus® was used as standard carrier matrix [54]. The reported study also determined the possibilities of up-scaling the process from a lab scale to a production line scale for industrial manufacturing. In that study, in order to save development time and material, the opportunity of predictability of the scale-up step was determined with a design of experiments (DoE) approach.

Soluplus® was extruded using three different sizes of co-rotating twin-screw extruders manufactured by Thermo Fisher (Germany), in different process settings following a DoE plan. As important process parameters, the RTD was measured with a tracer in each setup and the SMEC was calculated. Beside these special process parameters, all standard parameters e.g., $T_m$ at the extruder die, pressure at the die and torque were also measured. From the RTD, the mean residence time was calculated. RTD was obtained by measuring the concentration of a colour pigment with a photometric and a colorimetric method.

The data from three independent design-of-experiments were analysed using an analysis of variance and the resulting multi-dimensional regression models where used to calculate the design spaces which are compared for their overlap between the different scales of the extruders.

*Twin-screw extruders*: Three different sizes of parallel twin-screw extruder were used to simulate the scalability of the HME process: As a lab scale extruder, a Pharma 11, for medium scale, a Pharma 16 and for production scale, a Process 24 (Thermo Fisher Scientific, Germany) were used. All barrels were of a length of 40 L/D. The settings were varied to a minimum, mid-point and maximum value for the screw speed (100, 300 and 500 rpm), the temperature program (130, 165 and 200 °C) and the feed-rate were adjusted as follows: for 11 mm extruder 0.17–2.4 kg/h, for 16 mm 0.5–7.5 kg/h and for 24 mm 1.13–12 kg/h.

*Residence time*: In order to calculate residence time accurately, a pigment can be added as a tracer to the hopper of the feeding section at a given time $T_0$. The colour concentration can therefore be measured at the die over the time. There are two methods that are normally used to calculate the residence time accurately: the picture method where a picture of the strand is taken every 0.2 seconds and on every picture a defined size of strand is detected regarding the amount of red pixel; and secondly, the infrared method. ExtruVis 2 is a colorimeter, developed by Gryczke. It is used for measuring in-line the concentration of the pigment in the melt at the die exit [54].

*Results from a scale-up study*: To have a successful scale-up, it is necessary to have the same outputs and conditions for the material on the lab scale extruder as well as on the bigger, production scale extruder. Therefore, it is assumed that the residence time of the material within the extruder must be the same to allow melting and mixing on one hand and to avoid degradation on the other.

For the scale-up experiments, firstly the feed-rate was calculated by the equation of Schuler. The feed-rate was increasing from 1 to 3 kg/h according to this equation when changing from an 11 mm screw diameter to an extruder with 16 mm. When increasing the feed-rate by Schuler, the RTD on the next scale extruder was observed to be quite similar. Nevertheless, the distribution was observed to be narrower than the distribution of the lab scale extruder and slightly shorter as well. It was found that the RTD was a perfect match in terms of the SMEC [55]. Numerically, the SMEC is calculated by the torque, the screw speed and the feed-rate, while the screw speed and the throughput are predefined parameters (can be set individually) and the torque is a resulting value. Therefore the SMEC can be adjusted by adjusting the feed-rate.

For the scale-up from the design window of the Pharma 11 to the Pharma 16, only the feed-rate was adjusted to the bigger size. In the case of the scale-up to a 24 mm system, the feed-rate as well as the screw speeds were increased. Another effect which could be shown during successful scale-up exploitation *via* HME devices was the correlation of the SMEC with the degree of filling of the extruder. All these effects can be explained by the equation of the SMEC (**Equation 1.4**). With increasing feed-rate and therefore increasing volume-specific feed load (VSFL) there is a decreasing mechanical energy input, because more materials share the mechanical energy which is supplied by the system (**Figure 1.3**). Another point which is also very important is that with increasing barrel temperatures, the SMEC is decreasing. Actually, with increasing barrel temperature, the viscosity of the material will decrease and the torque will also follow the latter.

$$SEMEC = \frac{\tau . \eta}{m} \quad [\frac{kJ}{Kg}]$$

(1.4)

where $\tau$ is torque, $n$ is screw speed and $m$ is throughput.

With increasing feed-rate and increasing VSFL, the mechanical energy input will decrease, as more portions of the material will share the mechanical energy which is supplied by the system. Another point which is also very important is that with increasing barrel temperatures the SMEC will be decreasing. Similarly, with increasing barrel temperature the viscosity of the material will decrease, as does the torque (**Figure 1.3**).

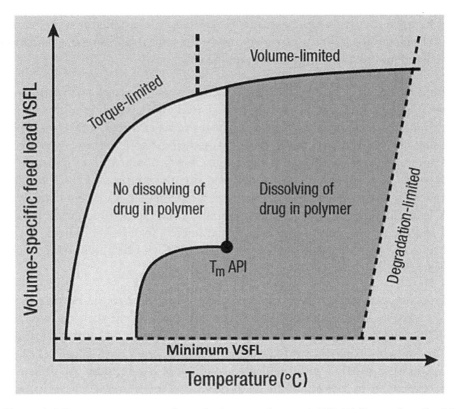

**Figure 1.3** Process parameter chart during a scale-up *via* HME. Reproduced with permission from K. Paulsen, D. Leister and A. Gryzcke in *Investigating Process Parameter Mechanism for Successful Scale-up of a Hot-Melt Extrusion Process*, 2013, Thermo Fisher Scientific, Note LR-71. ©2013, Thermo Fisher Scientific and BASF [54]

## 1.4 Conclusions

HME has proven to be an emerging technique to develop, optimise and engineer numerous drug delivery systems and therefore it has been found to be useful in the pharmaceutical industry enlarging the scope to include a range of polymers and API that can be processed. Being a solvent-free, robust, quick and economy favoured manufacturing process, HME has been exploited for the production of a large variety of pharmaceutical dosage forms. HME can also be adopted as a continuous manufacturing process, eliminating intermediate steps and higher labour costs. An appropriate use of HME in continuous manufacturing sphere is seen to be capable of revolutionising the pharmaceutical industries. Selection of suitable processing parameters and carefully optimised conditions may prove HME as a novel processing technique in successful and effective scale-up for industrial manufacturing purposes.

## References

1. S.P. Jagtap, S.S. Jain, N. Dand, R.K. Jadhav and J.V. Kadam, *Scholars Research Library*, 2012, **4**, 1, 42.

2. S. Madan and S. Madan, *Asian Journal of Pharmaceutical Sciences*, 2012, **7**, 2, 123.

3. R. Chokshi and H. Zia, *Iranian Journal of Pharmaceutical Research*, 2004, **3**, 3.

4. S. Singhal, V.K. Lohar and V. Arora, *WebmedCentral Pharmaceutical Sciences*, 2011, **2**, 1, WMC001459.

5. C. De Brabander, C. Vervaet, L. Fiermans and J.P. Remon, *International Journal of Pharmaceutics*, 2000, **199**, 195.

6. C. De Brabander, C. Vervaet and J.P. Remon, *Journal of Controlled Release*, 2003, **89**, 235.

7. H.H. Grunhagen and O. Muller, *Pharmaceutical Manufacturing International*, 1995, **1**, 167.

8. K. Vithani, M. Maniruzzaman, S. Mostafa, Y. Cuppok and D. Douroumis in *Proceedings of the 39th Annual Meeting and Exposition of Controlled Release Society*, 15-18th July, Quebec, Canada, 2012.

9. B. Abrahamsson, M. Alpsten, B. Bake, U.E. Jonsson, M. Eriksson-Lepkowska and A. Larsson, *Journal of Controlled Release*, 1998, **52**, 301.

10. A. Gryczke, S. Schminke, M. Maniruzzaman, J. Beck and D. Douroumis, *Colloids and Surface B: Biointerfaces*, 2011, **86**, 2, 275.

11. J. Vaassena, J. Bartscherb and J. Breitkreutza, *International Journal of Pharmaceutics*, 2012, **429**, 99.

12. K. Woertz, C. Tissen, P. Kleinebudde and J. Breitkreutz, *Journal of Pharmaceutical and Biomedical Analysis*, 2010, **51**, 497.

13. M. Maniruzzaman, J.S. Boateng, M.J. Snowden and D. Douroumis, *International Scholarly Research Notices: Pharmaceutics*, 2012, Article ID:436763.

14. M.M. Crowley, F. Zhang, A.M. Repka, S. Thumma, S.B. Uphadhye, S.K. Battu, J.V. McGinity and C. Martin, *Drug Development and Industrial Pharmacy*, 2007, **33**, 909.

15. J. Breitenbach, *European Journal of Pharmaceutics and Biopharmaceutics*, 2002, **54**, 107.

16. *The Dynisco Extrusion Processors Handbook*, 1st Edition, Eds., T. Whelan and D. Dunning, London School of Polymer Technology, Polytechnic of North London, London, UK, 1988.

17. G.P. Andrews, D.N. Margetson, D.S. Jones, S.M. McAllister and O.A. Diak in *United Kingdom & Ireland Controlled Release Society*, 2008, p.13.

18. M. Mollan in *Pharmaceutical Extrusion Technology*, Eds., I. Ghebre-Sellassie and C. Martin, CRC Press, Boca Raton, FL, USA, 2003, p.1.

19. A. Senouci, A. Smith and A. Richmond, *Chemical Engineering*, 1985, **417**, 30.

20. M. Repka, M. Munjal, M. Elsohly and S. Ross, *Drug Development and Industrial Pharmacy*, 2006, **32**, 21.

21. V.M. Litvinov, S. Guns, P. Adriaensens, B.J. Scholtens, M.P. Quaedflieg, R. Carleer and G. van den Mooter, *Molecular Pharmaceutics*, 2012, **9**, 10, 2924.

22. X. Zheng, R. Yang, X. Tang and L. Zheng, *Drug Development and Industrial Pharmacy*, 2007, **33**, 791.

23. J.O. Morales and J.T. McConville, *European Journal of Pharmaceutics and Biopharmaceutics*, 2011, **77**, 187.

24. N. Follonier, E. Doelker and E.T. Cole, *Journal of Controlled Release*, 1995, **36**, 342, 250.

25. F. Zhang and J.W. Mcginity, *Pharmaceutical Development and Technology*, 1998, **14**, 242.

26. Y. Miyagawa, T. Okabe, Y. Yamaguchi, M. Miyajima and H. Sunada, *International Journal of Pharmaceutics*, 1996, **138**, 215.

27. *Encyclopedia of Pharmaceutical Technology*, 3rd Edition, Ed., J. Swarbrick, Informa Healthcare, New York, NY, USA, 2006, p.1773.

28. M.A. Repka, T.G. Gerding, S.L. Repka and J.W. Mcginity, *Drug Development and Industrial Pharmacy*, 1999, **25**, 625.

29. N. Follonier, E. Doelker and E.T. Cole, *Drug Development and Industrial Pharmacy*, 1994, **20**, 1323.

30. G.P. Andrews, D.S. Jones and O.A. Diak, *European Journal Pharmaceutics and Biopharmaceutics*, 2008, **69**, 264.

31. J. Breitkreutz, F. El-Saleh, C. Kiera, P. Kleinebudde and W. Wiedey, *European Journal Pharmaceutics and Biopharmaceutics*, 2003, **56**, 255.

32. E. Roblegg, E. Jäger, A. Hodzic, G. Koscher, S. Mohr, A. Zimmer and J. Khinast, *European Journal of Pharmaceutics and Biopharmaceutics*, 2011, **79**, 635.

33. C.R. Young, C. Dietzsch, M. Cerea, M. Farrell, T. Fegely and K.A. Siahboomi, *International Journal of Pharmaceutics*, 2005, **301**, 112.

34. M. Wilson, M.A. Williams, D.S. Jones and G.P. Andrews, *Therapeutic Delivery*, 2012, **3**, 6, 787.

35. R. Singh, A. Sahay, K.M. Karry, F. Muzzio, M. Ierapetritou and R. Ramachandran, *International Journal of Pharmaceutics*, 2014, **473**, 1–2, 38.

36. M. Mezhericher, *European Journal of Pharmaceutics and Biopharmaceutics*, 2014, **88**, 866.

37. H. Patil, X. Feng, X. Ye, S. Majumdar and M.A. Repka, *AAPS Journal*, 2015, **17**, 194.

38. M. Maniruzzaman, M.M. Rana, J.S. Boateng, J.C. Mitchell and D. Douroumis, *Drug Development and Industrial Pharmacy*, 2013, **39**, 2, 218.

39. A.V. Trask and W. Jones, *Topics in Current Chemistry*, 2005, **254**, 41.

40. M.B. Hickey, M.L. Peterson, L.A. Scoppettuolo, S.L. Morrisette, A. Vetter, H. Guzman, J.F. Remenar, Z. Zhang, M.D. Tawa, S. Haley, M.J. Zaworotko and O. Almarsson, *European Journal of Pharmaceutics and Biopharmaceutics*, 2007, **67**, 1, 112.

41. D. McNamara, S. Childs, J. Giordano, A. Iarriccio, J. Cassidy, M. Shet, R. Mannion, E. O'Donnell and A. Park, *Pharmaceutical Research*, 2006, **23**, 8,1888.

42. V. Trask, W.D.S. Motherwell and W. Jones, *International Journal of Pharmaceutics*, 2006, **320**, 1–2, 114.

43. *Quality Guideline for Pharmaceutical Development, Q2 (R2)*, International Conference on Harmonisation, US Food and Drug Administration, Federal Register, 71 (106), 2006.

44. *Quality Guideline for Pharmaceutical Development, Q2 (R2)*, International Conference on Harmonisation, US Food and Drug Administration, Federal Register, 74 (66), 2009.

45. H. Wu and M.A. Khan, *Journal of Pharmaceutical Sciences*, 2010, **99**, 3, 1516.

46. L. Saerens, L. Dierickx, B. Lenain, C. Vervaet, J.P. Remon and T. De Beer, *European Journal of Pharmaceutics and Biopharmaceutics*, 2011, 77, 158.

47. R.S. Dhumal, A.L. Kelly, P. York, P.D. Coates and A. Paradkar, *Pharmaceutical Research*, 2010, **27**, 2725.

48. A.L. Kelly, T. Gough, R.S. Dhumal, S.A. Halsey and A. Paradkar, *International Journal of Pharmaceutics*, 2012, **426**, 1–2, 15.

49. J.F. Carley and J.M. Mckelvey, *Industrial and Engineering Chemistry Research*, 1953, **45**, 989.

50. C.I. Chung, *Polymer Engineering and Sciences*, 1984, **24**, 626.

51. H. Potente., *International Polymer Processing*, 1991, **6**, 267.

52. M. Nakatani, *Advances in Polymer Technology*, 1998, **17**, 19.

53. P.G. Andersen in *The Werner & Pfleiderer Twin-screw Co-rotating Extruder System: Plastics Compounding, Equipment and Processing*, Hanser Gardner Publications, Cincinnati, OH, USA, 1998.

54. K. Paulsen, D. Leister and A. Gryzcke in *Investigating Process Parameter Mechanism for Successful Scale-Up of a Hot-Melt Extrusion Process*, 2013, Thermo Scientific, Walton, MA, USA and BASF, Ludwigshafen, Germany, Note LR-71.

55. K. Kohlgrüber and C.H. Verlag, *Der Gleichläufige Doppelschneckenextruder: Grundlagen, Technologie, Anwendungen*, Hanser, Munich/Berlin, Germany, 2007.

# 2 Hot-Melt Extrusion as Continuous Manufacturing Technique, Current Trends and Future Perspectives

Ritesh Fule

## 2.1 Hot-Melt Extrusion as Continuous Manufacturing Technique

Innovative pharmaceutical manufacturing facility, based on continuous processing with in-line monitoring strategies such as process analytical tools, coupled with a resourceful programmed computational control system, is highly preferred for well-organised quality production of pharmaceutical formulations with respect to time, space, precision and resources. Currently, the pharmaceutical companies are facing massive formulation challenges to fulfill regulatory anticipations, commercial feasibility, functioning difficulties and economic restrictions. Because of patent related issues such as patent exclusivity, expiry and royalties, the generation of newly discovered drugs has been decreased considerably. The actual, life of a patent is less than 20 years even if it has 20 years of validity. Also, the high throughput method is useful for the development of a large number of drug candidates, but more than 60% of them have formulation related problems. So, formulation development is a core strategy for the commercial benefits of the pharmaceutical industry. Also, lower price competition with generic drug manufacturers after the expiry of patent exclusivity makes reductions in the cost of products sold to a major part of the population. Continuous manufacturing processes, which provide high efficiency, more desirable product quality, and optimal use of area, effort and resources, have evolved as an efficient and promising alternative for achieving these goals. Besides this, depending on active pharmaceutical ingredient(s) (API) and polymeric excipient properties for obtaining solid dosage quality product *via* different developing routes, use of manufacturing methods such as direct compaction, slugging, fluidised spray drying, roller compaction (dry granulation), or wet granulation will continue to be essential process requirements. Thus, with active product assortments demanding multiple manufacturing platforms, stationary singular manufacturing plants will not be sufficient. Batch processes are highly flexible and allow for multiple manufacturing routes. However, this flexibility comes at the high price of very low overall plant and equipment productivity. Therefore, manufacturing machinery which can provide both flexible and improved productivity is desired. Such a flexible multipurpose continuous tablet manufacturing process was developed at the Engineering Research Center for Structured Organic Particulate Synthesis (ERC-SOPS), based on the central concept of continuous processing. To enable the implementation of continuous manufacturing, extensive research has been performed to understand process dynamics using model-based dynamic flowsheet simulations as a design tool prior to planned implementation into the pilot-plant manufacturing facility as at the National Science Foundation's Engineering Research Center premises (Rutgers, USA) [1, 2]. The model-based virtual experimentation helps to optimise resources *via* reducing the number of real experiments, which is essential for the design and optimisation of a continuous process and the implementation of an efficient control system. The pharmaceutical industry is still dominated by 'stationary' batch processes with little, if any, automatic feedback control. This paradigm is predicated on the mindset that uniform products can only be produced by batch processes having a 'frozen' manufacturing process [3, 4].

Implementation of hot-melt extrusion (HME) as continuous manufacturing technology for the development of a variety of solid dosage formulation has enormous opportunities to overcome drug related problems. HME has been established as a novel strategy to produce a delivery system with enhanced bioavailability as well as solubility with improved dissolution rate, which is a focus of API research in pharmaceutical drug development. This technology employs the combination of optimised parameter and temperature to formulate drug-polymer molecularly dispersed systems, which can be termed as solid dispersion or solid solution. The singularity and workability of the procedural features allows the development of several drug delivery systems. Also, various marketed formulations are available prepared *via* HME [5–8]. The formation of melt extrudate involves the exchange of heat energy during HME process, followed by instant cooling of the melt, which affects the thermodynamic and kinetic properties involved in the forming of solid dispersions [9–11]. The use of a highly water-soluble carrier in solid dispersion always increases the chances of crystallisation due to swelling behaviour when it comes into contact with the aqueous gastrointestinal fluid [12]. Therefore, surface active agents or surfactants are used as inhibitors for recrystallisation. HME has the unique property to maintain the amorphous state of the drug after the formation of solid dispersion and improves the solubility of drug. Solubility of the drug substance has a significant impact on its bioavailability performance. This perspective also presumes a dated interpretation of regulatory quality requirements, where the process is 'validated' and never changed afterwards, missing the reality of continuously changing raw materials and other process inputs [13].

Herein, we present the first example of an end-to-end, united continuous manufacturing flowsheet for a pharmaceutical HME product. Our flowsheet starts from a pre-optimised ratio of API, polymeric excipients and lubricants and all the intermediate process, drying, and formulation, which results in a formed final tablet in a unique tightly, controlled process. This provides a platform to test newly developed continuous technologies within the context of a fully integrated production system, and to investigate the system-wide performance of multiple interconnected units. Herein, the key features of operating the system for durations of up to 1-day are presented. The one-day period included start-up of the plant, stabilisation of essential parameters, and periods of entire operational process. We specifically highlight areas which need implementation of advanced monitoring of continuous-flow features, and appropriate techniques applicable to continuous manufacturing of pharmaceutical product [14].

### 2.1.1 Hot-Melt Extrusion Equipment

The various types of extruders have a common feature of forcing the extrudate from a wider cross-section through the restriction of the die. The theoretical approach to understand the system is therefore, generally associated with dividing the process of flow into 4 sections:

1.  Feeding of the extruder,

2.  Conveying of mass and entry into the die,

3.  Flow through the die, and

4.  Exit from the die and downstream processing.

The four illustrated sections suggest considerations of different aspects such as flow of powder, shear force, residence time and pressure, cooling rate and shaping. A diagram of a typical extruder is shown in **Figure 2.1**. Normally, the extruder features a distinct rotating screw inside a stationary cylindrical barrel. The barrel is often manufactured in sections, which are bolted or clamped together. An end-plate die, connected to the end of the barrel, defines the shape of the extruded material. Additional methods include mass flow feeders to accurately meter materials into the feed hopper, process analytical tools to measure extrudate properties [near-infrared (NIR) systems and laser systems], liquid and solid side stuffers, and vacuum pumps to dolomitise extrudates, pelletisers and calendering equipment. Standard process control and monitoring strategies include zonal hotness and screw rapidity with voluntary monitoring of torque, drive amperage, and pressure and melt viscosity [15]. Temperatures are normally controlled by electrical heating bands and monitored by thermocouples.

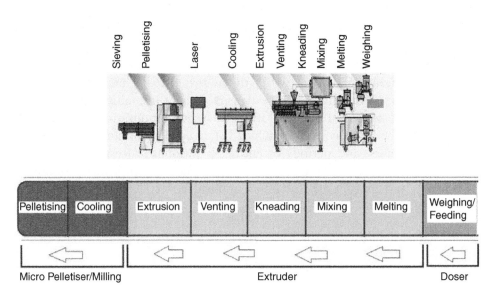

**Figure 2.1** A typical diagram of extruder design

The constituents inside the barrel are heated chiefly by the heat generated as a result of the shearing upshot of the rotating screw and the temperature conducted from the hot barrel. Occasionally as much as 80% of the heat prerequisite to liquefy or melt the constituents is supplied by the heat generated by friction. Additional heat may be supplied by electric or liquid heaters mounted on the barrel. It is essential to recognise that residence time and pressure in the die area could have a substantial influence on the impurity profile of the final extruded product. The molten mass is eventually pumped through the die which is attached to the end of the barrel. The extrudates are then subjected to further processing by auxiliary downstream devices. During a continuous extrusion process, the feedstock is required to have good flow properties inside the hopper. For the material to demonstrate good flow, the angle between the side wall of the feeding hopper and a horizontal line needs to be larger than the angle of repose of the feedstock. In the case of cohesive materials or for very fine powders, the feedstock tends to form a solid bridge at the throat of the hopper, resulting in erratic powder flow. For these situations, a force-feeding device is sometime used. Most screws are made from stainless steel that is surface-coated to withstand friction and potential surface erosion and any decay that may occur during the extrusion process [16].

The starting material is fed from a hopper directly into the feed section, which has deeper trips or grooves with pitch. This geometry enables the feed material to fall easily into the screw for conveying along the barrel. Pitch and helix angle determine the throughput at a constant rotation speed of the screws. The performance of the feeding section depends on the external friction coefficient of the feedstock at the surface of the screw and barrel. The material is transported as a solid plug to the transition zone where it is mixed, compressed, melted and plasticised. Compression is developed by decreasing the thread pitch but maintaining a constant flight depth or by decreasing flight depth while maintaining a constant thread pitch. Both methods result in increased pressure as the material moves along the barrel [17]. The melt moves by circulation in a helical path by means of transverse flow, drag flow, pressure flow, and leakage; the latter two mechanisms reverse the flow of material along the barrel. The space between the screw diameter and the width of the barrel is normally in the range of 0.1–0.2 mm. The friction at the inner surface of the barrel is the driving force for the material feed, whereas the friction at the surface of the screw restricts the forward motion of the material. A high friction coefficient in the barrel and a low friction coefficient at the screw surface would contribute to a more efficient transfer of the materials in the feed section. Other properties of the feedstock, such as bulk density, particle size, particle shape, and material compatibility, can also have an impact on the performance of the feeding section. The transfer of the material should be efficient in order to maintain an increase in pressure in the compression zone and the metering zone. The pressure rise in these zones should be high enough to provide an efficient output rate of the extrudate. It is also possible to fine tune the barrel temperature at the feeding section in order to optimise the friction at the surface of the barrel. Inconsistent material feed may result in a 'surge' phenomenon that will cause cyclical variations in the output rate, head pressure, and product quality [18]. The material reaches the metering zone in the form of a homogeneous plastic melt suitable for extrusion. For an extrudate of uniform thickness, flow must be consistent and without stagnant zones right up to the die entrance. The function of the metering zone is to reduce pulsating flow and ensure a uniform delivery rate through the die cavity [19].

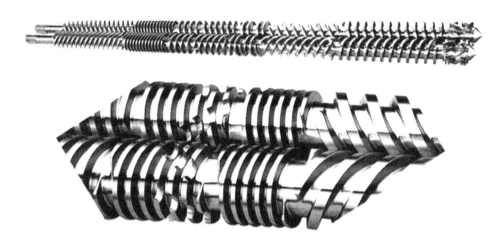

Figure 2.2 Represents a diagram of screw geometry

The die is attached to the end of the barrel. The geometrical design of the die will control the physical shape of the molten extrudate. The cross-section of the extrudate increases due to swelling as the molten mass exits the die. This is referred to as 'die swelling'. This entropy- driven event occurs when individual polymer chains recover from deformation imposed by the rotating screw by 'relaxing' and increasing their radius of gyration. Most commercial extruders have a modular design to facilitate changing screws [20]. The design of the screw has a significant impact on the process and can be selected to meet particular requirements such as high or low shear [21]. Specific screw features are displayed in **Figure 2.2**.

In an extrusion process, the dimensions of the screws are given in terms of the length and diameter (L/D) ratio, which is the length of the screw divided by the diameter. For example, an extruder screw that is 1,000 mm long and has a 25 mm diameter exhibits a 40:1 L/D. Typical extrusion process lengths are in the 20 to 40:1 L/D range, or longer. Extruder residence times are generally between 5 sec and 10 min, depending upon the L/D ratio, type of extruder, screw design, and how it is operated. The size of an extruder is generally described, based on the diameter of the screw used in the system, i.e., 18–27 mm extruder (pilot scale) as compared with 60 mm extruder (production scale). Although the screw size difference appears small (~2-fold) in the preceding example, the extruder output that results from doubling the screw size may be 10-fold, i.e., from 10 to 100 kg/h. This is due to the much larger volume available for processing as the screw size is increased [22].

### 2.1.2 Single-screw Extruder

The single-screw extruder is the most widely employed in manufacturing process developments. One screw rotates inside the barrel and is used for feeding, melting, devolatilising and pumping. Mixing is also accomplished for less demanding applications. There are three basic functions of a single-screw extruder: solids conveying, melting and pumping. The forwarding of the solid particles in the primary segment of the screw is an outcome of friction between the material and the feed section's bore. After solids conveying, the flight depth begins to taper down and the heated barrel causes a melt to form. The energy from the electric heater and shearing contribute to melting. Preferably, the melt pool will increase as the solid bed reduces in size until all is molten at the end of the compression zone. Finally, the molten materials are propelled against the die resistance to form the extrudate.

### 2.1.3 Twin-screw Extruders

The first twin-screw extruders were developed in the late 1930s in Italy, with the concept of combining the machine actions of several available devices into a single unit. As the name implies, twin-screw extruders utilise two screws usually arranged side by side (**Figure 2.3**) [23]. The use of two screws allows a number of different configurations to be obtained and imposes different conditions on all zones of the extruder, from the transfer of material from the hopper to the screw, all the way to the metered pumping zone. In a twin-screw extruder, the screws can either rotate in the same (co-rotating extruder) or the opposite (counter-rotating extruder) direction. The counter-rotating designs are utilised when very high shear regions are needed as they subject materials to very high shear forces as the material is squeezed through the gap between the two screws as they come together. Also, the extruder layout is good for dispersing particles in a blend. Generally, counter-rotating twin-screw extruders suffer from disadvantages of potential air entrapment, high-pressure generation, and low maximum screw speeds and output. Co-rotating twin-screw extruders on the other hand, are generally of the intermeshing design, and are thus self-wiping. They are industrially the most important type of extruders and can be operated at high screw speeds and achieve high outputs, while maintaining good mixing and conveying characteristics. Unlike counter-rotating extruders, they generally experience lower screw and barrel wear as they do not experience the outward 'pushing' effect due to screw rotation. These two primary types can be further classified as non-intermeshing and fully intermeshing. The fully intermeshing type of screw design is the most popular type used for twin-screw extruders [24].

**Figure 2.3** A twin-screw extruder diagram

### 2.1.4 Role of Glass Transition Temperature

There is no glass transition temperature ($T_g$), there is a glass transition region. The change from the glassy state into a liquid or a rubbery state is gradual [25]. $T_g$ values are reported by analogy with the melting temperature ($T_m$) values, so as to represent a region by a single number. While $T_m$ values do not depend on the direction of the change (freezing a liquid, melting a solid) or on the change rate, the location of the glass transition region depends on both factors. In his classical 1958 study of polyvinyl acetate Kovacs [26] has shown how the $T_g$ location varies with the cooling rate of the liquid. Full miscibility is characterised by a single glass transition temperature for all the blends [27]. In immiscible polymers, not an infrequent case, $T_g$ values for pure components do not change with composition. The miscibility (or lack of it) is decisive for all properties [28].

### 2.1.5 Role of Mechanical Parameters

Mechanical parameters such as screw configuration, rotating speed, die size and diameters also influenced the quality of extrudates. Depending upon the types of die, different dosage forms such as granules, pellets, spheres, films, transdermal patches and implant based formulation systems can be manufactured. The prescreening of these parameters will be carried out by considering the extrudate uniformity, yield, the homogeneity of the formed extrudate and also by using design software tools for early optimisation [29].

### 2.1.6 Use of Plasticisers or Surfactants

The use of plasticisers, surfactants or inorganic excipients has been reported by many researchers as having additional advantages for an improved HME process. The use of polyethylene glycol, grades of poloxamers, D-α-tocopheryl polyethylene glycol succinate, triethyl citrate, amino acids, and methacrylates has been incorporated in sufficient amounts to make the HME process easier and to improve the quality of the extrudates. The increase in wettability, reduction in melt viscosity, reduction of torque, reduction of pressure and faster processing are the fundamental advantages of additional plasticising agents. In certain processes, the use of a polymer alone is not sufficient to reduce the $T_g$ of a drug and solubilise it inside a polymer matrix. Thus, surfactants help to move the polymeric chains at lower $T_g$ and thereby improve the HME process performance [30–32].

### 2.1.7 Challenges and Opportunities

The pharmaceutical industry has reached 1 trillion USD with its large, high value added manufacturing sector and increasing annual worldwide sales. The traditional manufacturing mode in this sector has been batch operation with staggered processes such as research and development at lab scale, technology transfer at pilot scale and large scale production of final dosage form. These stationary processes increase the cost and labour of the development of a particular product. Also, documentation for regulatory compliance will require brainstorming as one has to develop it by visiting several sites of manufacturing. But, everything has been set as per the US Food and Drug Administration (FDA), European Medicines Agency and Medicines and Healthcare Products Regulatory guidelines, and pharmaceutical companies are following those guidelines. So, although it's essential, from the manufacturer's point-of-view, it is still not clear whether such types of continuous processes with monitoring tools will be approved by the regulatory bodies or not. However, recent advances in technologies, changes in the regulatory climate and continuous drivers for cost reduction have provided a unique opportunity for the introduction of advanced manufacturing technologies.

Continuous processing is considered to be one of the crucial technologies that can provide substantial innovation in the pharmaceutical segment, also motivated by the vision of developing 'in need' personalised medicines. In addition to offering better product consistency and overall process efficiency, continuous manufacturing has the potential to provide more distributed and even mobile manufacturing systems that could be located at the point of use, improving access to novel medicines, and opening up new markets. HME as a continuous manufacturing technique has emerged as having immense potential to develop value added pharmaceutical formulation with commercial feasibility.

HME provides the opportunity to reduce costs, drive innovation and accelerate the time to market. However, to be able to exploit the advantages of continuous manufacturing processes in an industry characterised by extraordinary value, a high variety and high volume of products, obtained through a network of distributed manufacturing systems is required, together with advances in fundamental process understanding, continuous processing and equipment particularly for chemical solids and in- measurement, modelling and control methodologies.

## 2.2 Background Information on Batch Processing

In reality, batch processes are not particularly good for product quality assurance, and possess a number of drawbacks such as poor controllability, low yield, problematic scalability, and energy inefficiency. In addition, batch processes are labour intensive and typically exhibit low plant productivity. As mentioned, one of the advantages of the batch processes is flexibility. Thus, ongoing efforts have focused on designing significant flexibility into continuous processes, to counter this 'advantage' of batch systems. Given the unique convergence of the patent bluff and the growing quality expectations of the regulatory system, the time is now right for the pharmaceutical industry to evolve from the 'frozen' batch manufacturing process characteristic of the 19th century, to automatic controlled continuous manufacturing which will dominate the 21st century [33, 34].

Continuous manufacturing, computerisation and regulatory compliance, including advanced model predictive control (MPC), are known to industry, and have been widely used in processing industries such as oil refining and commodity chemicals [35]. As a closed circlet optimal control method, based on the categorical use of a process model, MPC has proven to be a very effective controller design strategy over the last twenty years. However, in the pharmaceutical industry, the application of continuous manufacturing, automation and control is still a challenging task [36]. Partly, this is due to the high level of design and operational complexity involved such as:

• Powders and granules do not flow as competently as fluids, and this is authoritative as material flow is a principal distinguishing fact of continuous manufacturing.

• Solid-state processing developments, such as drying and coating, can require more residence times, thus requiring also special strategies to accommodate this.

• Rheology, physical properties and thermal behaviour of drug-polymer.

• The output of one unit operation does not inevitably equal exactly the input to the subsequent unit operation, therefore surge capacity hoppers might be needed to connect some unit operations and to act as buffers.

• Inaccessibility of appropriate on-line monitoring tools and involuntary feedback control systems in distinct system constituents also make continuous operation difficult.

### 2.2.1 Past and Future Trends of Continuous Manufacturing

Over the last few years the pharmaceutical industry has proved to be one of the booming areas of development after information technology. Due to the advantages of HME over conventional solid dosage form manufacturing techniques, the pharmaceutical industry has become interested in this innovative technology. Besides the continuity of the process, its major advantages are fewer processing steps, no use of organic solvents or water and the possibility of improving drug solubility or sustaining drug release. Additional benefits of this technique include its versatility, increased throughput and reduced costs. By producing multilayer products with a reduced amount of expensive polymers and an increased amount of inexpensive polymers, a cost-efficient process can be achieved without sacrificing performance [38, 39]. The technology does have a price for initial set up. This includes investment in equipment (additional extruders) and the need for additional floor space for the extruders, as well as the need for an experienced line operator (taking into account the increased levels of process complexity). In some cases the additional process costs may offset the material cost savings [40].

### 2.2.2 Conventional Batch Process Design

The conventional batch process design is time consuming with more frozen processes requiring more labour and resources. The conventional batch process requires monitoring at each level and it decreases overall productivity and effectiveness. But, the regulatory guidelines must need to be followed which are developed for batch processing.

### 2.2.3 Production Parameters

From bench scale to pilot scale, entire documentation and analysis reports must comply with the regulatory guidelines for respective dosage forms. The drug master file (DMF) for the production batch includes all the data relating to the API supplier, maintenance, good manufacturing practice of manufacturing facility and good laboratory practice measures.

### 2.2.4 Production Parameters in Batch Processing

#### 2.2.4.1 Active Pharmaceutical Ingredient(s) Manufacturing Site

The API manufacturer has to synthesise the drug product under strict regulatory and environmentally friendly conditions. The synthesised API with all the physical and chemical characteristics should be documented with purity profiles during submission of the DMF. The manufacturing process and maintenance of the final synthesised product should be included in the DMF. Transportation of API to the formulation unit is also an essential part of the batch process.

#### 2.2.4.2 Formulation Development

Formulation of a drug into the desired dosage form depends upon the characteristics of the drug and the additives with respect to the desired delivery system. The detailed pre-formulation, optimisation and development of the formulation at lab scale was developed and then the pilot batch was reproduced on an industrial scale in triplicate using various API from the same supplier. Qualitative validation and quantitative evaluation needs to be carried out in order to calculate the similarities and error of variance to confirm the effectiveness of final product.

### 2.2.4.3 Industrial Scale-up

The industrial scale-up process must be robust, continuous and errorless. The convictive and continuous process during scale-up may suffer from certain issues which need to be resolved with modification.

### 2.2.4.4 Regulatory Guidelines

One has to follow the regulatory guidelines for different dosage form manufacturing to improve and sustain the quality attribute of pharmaceutical formulation. There are codes of federal regulation guidelines published by the FDA and other agencies to control the production of the pharmaceutical healthcare industry. CFR21 is a specific Code of Federal Regulations which involves various guidelines with respect to different dosage regimens [41].

### 2.2.5 Validation of Batch Processing

Process validation is defined as the collection and evaluation of data, from the process design stage through commercial production, which establishes scientific evidence that a process is capable of consistently delivering quality product.

This guidance aligns process validation activities with a product lifecycle concept and with existing FDA guidance, including the FDA/International Conference on Harmonisation Guidance for Industry, Q8 (R2) Pharmaceutical Development, Q9 Quality Risk Management, and Q10 Pharmaceutical Quality System. Although this guidance does not repeat the concepts and principles explained in the above, FDA encourages the use of modern pharmaceutical development concepts, quality risk management, and quality systems at all stages of the manufacturing process lifecycle. This guidance incorporates principles and approaches that all manufacturers can use to validate manufacturing processes [42].

### 2.2.6 In-line Continuous Monitoring Techniques

In HME, the in process analytical tools include the use of in-line spectroscopy such as NIR and Raman for multicomponent compositional monitoring. In process development and scale-up, the spectroscopic in-line compositional methods are used for determination of residence time distributions, process description, and system documentation [43]. In a manufacturing setting these tools are used for continuous verification that the process is producing material at the target composition, and for real time isolation of off-specification product due to disturbances from the feeding systems. Commercial extruders incorporate functionalities such as melt supervisory temperature control, die pressure, feed-rate control and torque sensors. These measurements together with the use of spectroscopic procedures make it possible to provide multivariate regulatory monitoring for accountability recognition. Additionally, in-line/on-line visible spectroscopy has the potential for monitoring degradation and extrudate colour, if colour matching is important for the product [44]. On-line NIR spectroscopy is used consistently in the industry for understanding blending operations. NIR, mounted on the lid of the blender, can be used to collect several spectra at each revolution. Both qualitative and quantitative methods can be developed to assess blend uniformity. Application of qualitative methods is the more common approach due to the simplicity. In qualitative methods, measurement of spectral variability is pursued as a utility to quantify the extrusion process. Applications of Raman and/or NIR spectroscopy for the in-process monitoring of pharmaceutical production processes are shown in **Table 2.1** [45, 46].

Collective measures are relative standard deviation, API peak height, and API peak area. Blend consistency is attained when the spectral changeability magnitude touches a sustained minimum. Quantitative approaches associate blend composition to NIR spectra with PLS models. Other analytical techniques such as Raman and Laser induced fluorescence spectroscopy have appeared in the literature and the market place, but have not expanded to receive wide industry approval. The same methodologies have been applied to blend and granule lubrication uniformity, although lubricant compositional uniformity determined by in-line NIR does not directly assess the extent of lubrication [47].

Multivariate process monitoring predictive monitoring techniques and predictive analytics software packages.

Strategic process monitoring is a communal tool in the pharmaceutical engineering industry. Customary methodologies comprise univariate statistical process control charting, with the application rules for common cause variation. The critical process parameters and product quality attributes have been studied intrinsically using univariate control charting methodology [48]. Furthermore, conventional descriptive statistics are used to provide functional historical data for entire process trains, on a batch-to-batch basis. Process monitoring is conducted primarily for two reasons: as a source of authentication, that the process is running within the parameter space allowed by the regulatory filing, and for the development of process

**Table 2.1 Applications of Raman and/or NIR spectroscopy for the in-process monitoring of pharmaceutical production processes**

| Types of dosage form | Formulation operation | Raman/ NIR | In-process interfacing | Monitored critical process information | Evaluated challenges | Data-analyis method | Spectral preprocessing method |
|---|---|---|---|---|---|---|---|
| Pelletisation | Monitoring of pelletisation | Raman/ NIR | At-line | API solid-state changes | Complementary Raman and NIR | Univariate | 2nd Derivatives |
| | | NIR | | Process-induced transformations | | | 2nd Derivatives + Savitzky-Golay smoothing |
| Coating | In-line monitoring of fluid bed coating | NIR | Invasive | Prediction of film thickness | Dynamic calibration | PCA  PLS | MSC |
| | Monitoring of pan coating | Raman | At-line | Determination of coat thickness | Presence of strong fluorescent interference | Univariate + multivariate | SNV  MSC  2nd Derivatives Savitzky-Golay smoothing |
| | In-line monitoring of pan coating | | Non-invasive. To protect the probe against dust, compressed air was blown through an iron pipe, which was attached in front of the probe | Quantitative determination of API in coat + end-point determination | Active coating process | PLS | SNV |

| Process | Description | Technique | Sampling | Application | Purpose | Chemometrics | Preprocessing |
|---|---|---|---|---|---|---|---|
| Tablet | Monitoring of tablet manufacturing | NIR | Non-invasive | Content uniformity, compression force, crushing strength | Use of rapid content NIR analyser | PLS | SNV + MSC + Savitzky-Golay 1st derivatives |
| Fluid bed drying | In-line monitoring of fluid bed drying | NIR and Raman | Non-invasive | API phase transformations during drying | Quantification | PLS | SNV |
| Freeze-drying | In-line monitoring of lyophilisation | Raman/NIR | Non-contact probes built in the freeze-drier chamber | Product monitoring and process phase end-point determination | Raman *versus* NIR | PCA | Pearson's baseline correction |
| | | Raman | | In-line product behaviour monitoring | Controlled crystallisation of product | | |
| HME | In-line monitoring of HME | Raman | Contact probe built in the die | In-line product behaviour monitoring | API quantification + polymer/drug interactions | PCA / PLS | Savitzky-Golay and SNV |
| | | NIR | | | Polymer-drug quantification and solid-state characterisation | | SNV |

MSC: Multiplicative scatter correction

PCA: Principal component analysis

PLS: Partial least squares

SNV: Standard normal variate

knowledge/understanding. Additional motivations for process monitoring include preventative actions such as fault detection, and for process control, such as end point determination. Over the last 10 years, the use of multivariate statistical process control charting has emerged within the industry [49].

Several commercial real time multivariate process monitoring software packages that are also suitable for continuous processes are currently available. **Tables 2.2** and **2.3** present some examples of commercially available run time multivariate process monitoring technology products. Most of these products listed utilise latent variable methods (PCA, PLS), and descriptive statistics, but other types of analysis are possible, such as neural networks, cluster analysis, and tree methods.

| Table 2.2 Represents software packages available | |
| --- | --- |
| Company name | Product name |
| Umetrics | SIMCA4000, SIMCA Batch On-line |
| ProSensus | ProSensus Online |
| Unscrambler | Process Pulse |
| Ge-Fanuc | Proficy Cause+ , Troubleshooter |
| Stat Soft | Statistica Data Miner |

| Table 2.3 Represents company name and functionality of software packages | | |
| --- | --- | --- |
| Company name | Product name | Functionality |
| Siemens | SIPAT | PAT specific |
| Optimal | SynTQ | PAT specific |
| Ge-Fanuc | Proficy | Predictive analytics |
| Emerson | Plant Web | Predictive analytics |
| Aegis | Nexus | Predictive analytics |

Multivariate process monitoring is conducted in real time on the individual unit operations level, and models for successive process steps are easily strung together. The data for the entire process is typically analysed after batch completion. In addition to the primary process equipment, data from supporting equipment systems (feed tanks, steam generation and so on) can also be included in monitoring schemes. Aggregated data from all systems, over multiple production campaigns are holistically analysed off-line to detect trends, process drifts, and to develop correlations. This holistic process analysis can be automated with the use of plant-wide information technology systems.

Real time multivariate process monitoring is applicable to both continuous and batch processes, with the latter being the more common application in the pharmaceutical industry. These tools are applied to batch processes in the pharmaceutical industry such as high shear wet granulation and fluid bed drying [50–53]. Although these tools are often deployed only for process monitoring and fault detection, they can also provide run time predictive analytics capabilities.

## 2.3 Assimilation of Complete Quality-by-Design Models

This part of the chapter will provide an overview of the advantages and challenges, including regulatory aspects, relating to the continuous manufacturing of pharmaceuticals, from API feeding to formulation of the final product.

The inspiration is on resourceful use of on-line measurement, process modelling and control, excellent procedure management and real time process management. Process arrangements including the existing range of component operations from powder feeding to tablet coating to dissolution rate are under active consideration. The principles of process flowsheet modelling for modified processes are beginning to be established, offering the potential for design optimisation.

### 2.3.1 Chief Modification in Procedural Expertise for Continuous Manufacturing

In pharmaceuticals, continuous manufacturing is deliberately a cost effective approach but may not be useful for every processed material. Also, it is very difficult to develop the controlled continuous system whilst conforming to the regulatory guidelines. Still, today 80% of delivery systems are in the form of tablet dosage due to its compliance with regard to both patient administration and also commercial benefits. HME is a green processing technique which was introduced initially in the plastics industry. Since then, over the last three decades, HME has also been convincingly implemented within the pharmaceutical formulation industries [54].

In batch manufacturing, process drugs and intermediates are produced in batches which are individually tested and tracked. Testing is expensive and quality relies on product and intermediate testing instead of a process control system. A design of experiment (DoE) is required as the basis for early optimisation of drug excipient, flow property optimisation, cascade control strategy and modulated controlled system [55].

A combined quality-by-design (QbD) and Discrete Element Model simulation-approach is useful to evaluate a blending unit operation by estimating the impact of formulation parameters and process variables on the blending quality and blending end point. Understanding the unevenness of the API and the excipients, also their influence on the amalgamation progression, are critical essentials for blending QbD. In a first step, the QbD-methodology is systematically used to: 1) generate the critical quality attribute(s) (CQA) content uniformity and to link this CQA to its substitute blend homogeneity; 2) identify theoretically critical contributing elements that may influence blending procedure superiority; and 3) risk based analysis of these factors to outline the steps for process categorisation [56, 57].

## 2.3.2 Process Parameters for Preformulation for Hot-Melt Extrusion

In HME based formulation development, the selection of the drug and the prescreening of the polymers depend upon various evaluation factors such as calculation of solubility parameters, mathematical modelling, experimental design for optimisation of drug-polymer ratio, $T_g$ optimisation, in-line monitoring, quality of extrudate, physicochemical characterisation, drug-polymer miscibility, and so on.

Determination of the solubility parameter helps to understand the compatibility between selected polymer and API. The API and polymer have miscibility if the difference between the solubility parameter ($\Delta\delta$) is not more than equal to 7. The molecular weight and viscosity of the polymer also attributes to the miscibility. In some polymers the $\Delta\delta$ value passes but due to low viscosity and smaller surface area it would not give a uniformly extruded product. Thus, extrusion processing temperatures need to be optimised based on the polymorphic transformation and quality of the extrudates. In-line, at-line and off-line monitoring techniques such as NIR spectroscopy, Raman spectroscopy, atomic force microscopy, small-angle X-ray scattering, wide-angle X-ray diffraction, particle size distribution and hot stage microscopy have been successfully used by researchers to quantify for HME based formulations [58–60]. The multivariate analysis can also be one of the essential examples of analysing in-line test results.

## 2.3.3 Implementation of a New Product Development Process

### 2.3.3.1 Modelling Simulation of Hot-Melt Extrusion Process

Evaluation and understanding of the melt flow process could be an essential parameter for continuous HME based manufacturing process. Continual flow of melt formed from fed blend components during HME causes intermixing at a molecular level. The frictional forces, temperature, attrition impact between the wall of the extruder and screw play an important role in the formation of a uniform melt [61]. The screw configuration, helix angle, ratio of drug to polymer and die geometry, are essential factors which could be used in molecular simulation studies. Especially in the area of melt extrusion, in pharmaceuticals the literature does not teach any findings which focus on the flow of melt right from mixing to the cooling die section of the HME product. Physical, chemical and mechanical understanding of the melt flow process will improve the quality of product. Also, formation of pellets or spheronised particles or granule based formulation delivery system has acquired emerging interest as potential dosage form [62]. The HME processed material can be spheronised into different size particles as per desired product requirement.

The scale-up of the HME process requires a thorough understanding of the physics of extrusion mechanics and the chemistry of drug–polymer excipient interaction by various multivariate analytical methods. This melt flow study helps to understand how the simulation of the HME process in terms of die design and screw geometry could be optimised to produce a quality product. The mathematical calculation and procedural insights are explained with an example. The factors such as temperature distribution, melt flow, die size, melt viscosity in a lab scale and at production level have been explored. Also, the rotating extruder ridges and the angle geometry were studied. OpenFoam® simulation software is freely available which can be used for performing simulation operations.

The extruder parts are divided into hopper, feeder (feeding zone with regulator), mixing zone (20–40 °C), melting zone (50–100 °C), cooling zone (20–0 °C) and shaping zone (die). In addition, a conveyor belt helps to withdraw extruded material *via* the orifice at a constant rate. From the conveyor the extruded material goes in to the cutter mill where the extrudate is reshaped to the desired size of the granules. These granules can be filled into capsules or compressed into tablet dosage form. For the uniform mixed distribution of blend components, 1 the melt formation temperature, melt viscosity and flow are determining factors for the quality of extrudate. Extrusion temperature is most essential factor to form an homogeneous molecular solid dispersion or solid solution during HME. It is essential to maintain constant temperature conditions during the flow of the material through various parts such as the mixing, melting and cooling zones. Overall, the flow of the melt extrudate, melting temperature of extrusion, temperature attributed to intermixing, and the thermoplastic behaviour of the blend play crucial roles in the development of a quality extruded product [63].

Therefore, the quality designed based process rather than a trial and error method is required in order to produce a consistent product with the desired quality performance of a predetermined standard. A rational continuous flow design system can be produced which gives continuous product formation with a in process monitored process analytical technology (PAT) framework [64]. The shear thinning of melt depends mainly on the type of polymer and the extrusion temperature. It was also envisaged that the zero shear viscosity was strongly influenced by the ratio of drug to polymer. Hence, viscosity within melt can vary significantly as it influences the mixing of feed material. Also, transport, mixing and dispersive mixing are of central importance for the design of the screw section, and the design of the die impacts strongly on the quality, shape and uniformity of the pellets. Thus the appropriate design of the extrusion die is extremely important to achieve the desired shape and dimension of the extrudate [65]. Implementation of DoE, PAT and multivariate methods of analysis is used for pre-optimisation for both on-line and off-line operations during product development. Continuous manufacturing using HME is one of the challenging traits of pharmaceutical formulation development [66].

Disadvantages of batch manufacturing:

- Distinct batch size (output quantity driven by batch size)

- Several, consecutive process steps

- Many disruptions among/during process stages

- Lengthy waiting periods during lone process phases

- Frequent transportation periods during development phases

- Prolonged output periods from start to end

- High volumes of raw material and intermediate records

- Extensive qualitative validation and scale-up undertakings required

- Quality validated by in-process sampling and end-product testing

Advantages of continuous manufacturing:

- Quantifiable product quality throughout the production process

- Reduction of systems tracking (40–90%)

- Saving in capital costs (25–60%)

- Decrease in operational costs (25–60%)

- Reduction of raw material and intermediate inventories

- Fewer complex problems in scale-up

- Reduction in drug substance and drug product development periods

- Predetermined objectives facilitating time to market

Overall, despite constant monitoring and consistent processing there might be occurrence of errors during the industrial scale-up. Thus, batch manufacturing of formulations mostly takes 9–12 months for completion. There is, therefore, a need to develop innovative methods for the continuous manufacturing of pharmaceuticals [67].

### 2.3.4 Flow Chart Model for Continuous Hot-Melt Extrusion based Pharmaceutical Manufacturing Process

For the development of amorphous dispersions, melt extrusion is considered from an operational and cost-efficient standpoint, to be the chief technology used foremost by pharmaceutical corporations. The efficient processing of melt-extruded amorphous systems requires resilient knowledge of formulation and procedure to yield a system having the essential artifact qualities. As a result of the complexity associated with formulation investigation, an organised methodology for delivery formulation design is necessary to certify that major development standards should be fulfilled.

A flexible multipurpose continuous tablet manufacturing process, together with a simple control system [Proportional–Integral–Derivative (PID)], has been previously reported [68]. In this process, a typical pharmaceutical tablet can be produced through different paths, such as direct compaction (route 1), dry granulation using roller compaction (route 2) or wet granulation (route 3), and the most appropriate route is decided based on the properties of the raw materials and the desired characteristics of the final product. Route 1 is the simplest process while routes 2 and 3 provide enhanced flowability characteristics of the powder blends. Route 2 is particularly suitable for cases where powder feeds are moisture sensitive. The process is described in detail by Singh and co-workers [68] and has been previously extensively studied.

MPC is a more multifaceted technique that requires the development of higher conformity models and robust optimisation approaches. Moreover, depending on the process dynamics, an MPC scheme that requires a detailed process model and a computationally expensive optimisation strategy may not be obligatory, and on the other hand, a more computationally well-organised PID control scheme can be used. Therefore, a combined control stratagem integrating the benefits of both MPC and PID is particularly desirable for the structure-inclusive control of a progression comprised of fast and deferred dynamics. MPC has established itself in industry as an important advanced process control strategy [69], because of its advantages over conventional regulatory controllers [70], and it continues to be the technology of choice for constrained multivariable control applications in the process industry [71]. MPC refers to a family of control algorithms that employs an explicit model to predict the future behaviour of the process over an extended prediction horizon. These algorithms are formulated as a performance objective function, which is defined as a combination of set point tracking performance and control effort [72]. This objective function is minimised by computing a profile of controller output moves over a control horizon.

Mostly the HME pharmaceutical process used to manufacture solid dose systems that would require a cascade control strategy for this particular process. For that reason, in solid formulation operations, the cascade PID controller can be used with MPC as the supervisory controller. MPC is placed over the regulatory control level when planning the cascade control [73]. Depending on the process dynamics, MPC may also directly operate valve position signals rather than PID set points. For the simulation and development of a continuous tablet manufacturing plant prior to the final execution of the flexible plant, a flowsheet simulation platform has been developed within which all process knowledge comes together in order to allow the numerical solution of mass and energy balances and to simulate the variable process conditions at different stages of production. Flowsheet models have been described as the quantitative realisation of the operation of a chemical plant [74]. Also empirical approaches have been used for process specification and implementation. There are many challenges that need to be overcome in order to develop robust and reliable flowsheet models for solids processes. The model should include a description of all unit operations and an evaluation of their essential parameters, together with a description of operational and controlled variables and possible interactions, and finally, the integration of process design and control to identify globally valid operating conditions are essential for controlled quality based continuous HME processing [75].

## 2.3.5 Foremost Alteration in Structural Provisions for Quality and Technical Operations

Extensive research is presently being implemented to develop the predictive models for diverse complex unit operations involved in the continuous HME manufacturing process. A summary of some of these process models has been reported in literature [76]. The incorporation of numerous unit operation models in dynamic flowsheet platforms needs the interconnection of entire process streams and the identification of critical connecting properties which should be communicated across units. Through the simulation of different integrated process configurations, the evolution of key particle properties during the transient state (start-up and shutdown) is enabled. In addition, the effect of changes in process parameters and/or material properties which typically can vary during continuous manufacturing can be observed and analysed. Furthermore, through the implementation of various operating scenarios, the flowsheet model can be used for the comparison and assessment of different process alternatives, which can be used in scaling up to the desired plant size. Finally, the developed and validated flowsheet simulation system can be used for operator training, since any sequences in operating schedules can be virtually performed and analysed through a computer screen. As shown in **Figure 2.4**, there are three gravimetric feeders that provide the necessary lubricant, API and excipient to the system. In addition to feeders there is adjustment for the sieve which can be fixed as per particle size requirement. The feeders contain a hopper that can hold up to a certain amount of material and a rotating screw to change the flow rate. These feed streams are then connected to a blender within which an homogeneous powder mixture of all the ingredients is ideally generated. The physical mixture is then feed into the hopper of the HME assembly at an optimised rate. It is then processed *via* screw through various mixing, melting and cooling zones to transfer into the desired extruded shape system. Based on various die sizes, we can obtain different types of extrudates which can be cut to size by the cutter mill. If there are two or more API which are required to be formulated with different dissolution or release characteristics, a co-extrusion technique can be used in place of the commonly used HME instrument which is shown in **Figure 2.4**. The product flow is measured and passed directly into the tablet press for compression. Subsequently, the outlet from the blender is connected to a hopper to ensure a continuous flow of materials (appropriate amount) directly to the tablet press. Finally, tablets are obtained from the tablet press that could be sent for coating and a fraction of them can be tested for dissolution, hardness, and other properties. Then the properties such as friability and hardness are monitored by introducing a control strategy at this stage which is connected to MPC. The selected tablets are further evaluated for dissolution performance.

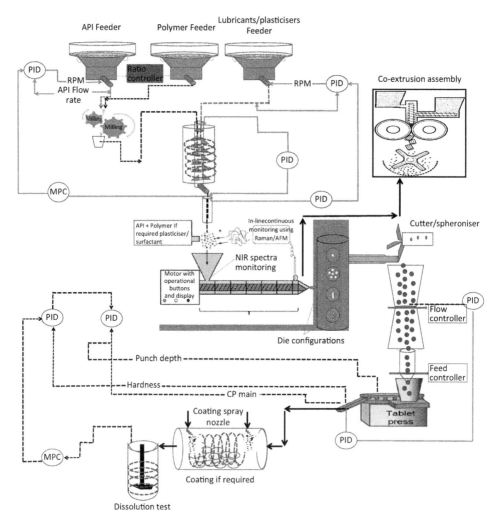

**Figure 2.4** Represents a continuous HME manufacturing flow chart with monitoring system

As literature teaches, solubility and dissolution rate are the two most influential properties of API in formulation development which have an effect on the therapeutic activity after administration of the desired dosage form. The literature cites various methods for preparing amorphous solid dispersions such as melt method, solvent evaporation, cyclodextrin inclusion complex and cryomilling which explains the importance of the solid dispersion type of formulation strategy [77]. Still, HME as a continuous process has advantages over the other conventional process. The nondestructive method helps to manufacture the product at less time with effective cost. There are products available in markets which are manufacture *via* HME. A list of some of the products under the approval process and marketing can be found in **Table 2.4** [78].

Table 2.4 Available HME based marketed and under development pharmaceutical products

| Innovator product | API content | Company | Category | HME formulation | Approval status |
|---|---|---|---|---|---|
| Zoladex | Bicalutamide | AsraZeneca | Prostate cancer | Shaped system | Marketed |
| Lacrisert | Hydroxypropyl cellulose | Merk | Dry eye syndrome | Shaped system | Marketed |
| Gris-PEG | Grisiofulvine | Pedinol Pharma | Antifungal | Crystalline dispersion | Marketed |
| Norvir | Ritonavir | Abbott Pharma | Antiviral (HIV) | Amorphous dispersion | Marketed |
| Implanon | Etonegestrel | Organon | Contraceptive | Implant system | Marketed |
| NuvaRing | Etonegestrel + Ethinyl Estradiol | Merck | Contraceptive | Implant system | Marketed |
| Eucreas | Vidagliptin + Metformin | Novartis | Diabetes | Melt granulation | Marketed |
| Kaletra | Lopinavir + Ritonavir | Abbott | Antiviral (HIV) | Amorphous dispersion | Marketed |
| Zithromax | Azithromycine | Pfizer | Antibiotic | Melt congealing | Marketed |
| Orzuedex | Dexamethasone | Allergen | Macular edema | Implant systems | Marketed |
| Fenoglide | Fenofibrate | LifeCycle Pharma | Dyslipidemia | MeltDose (solid dispersion) | Marketed |
| Posaconazole | Posaconazole | Merck | Antifungal | Amorphous dispersion | Development |
| Anacetrapib | Anacetrapib | Merck | Atherosclerosis | Amorphous dispersion | Development |

HIV: Human immunodeficiency virus

### 2.3.6 Potential Benefits of Continuous Hot-Melt Extrusion Manufacturing

Continuous manufacturing has attracted the attention of industry and academia alike by promising lower costs, greater reliability and safety, better sustainability, and novel pathways that are not otherwise accessible [79]. Recent studies have demonstrated that economic savings can be realised for certain cases by transforming a batch production into a continuous process. With existing batch-based manufacturing methods, it can take up to 12 months between the start of the first synthetic step and market release of the finished tablets [80], which partially is a consequence of movement of materials around and between facilities, and lengthy final product testing. This results in large and expensive inventories and shortages from manufacturing delays if the batch fails during the final testing once the production has finished. Continuous manufacturing allows a faster response to changes in demand; this permits a smaller inventory than for batch-based manufacturing, which not only results in lower working capital, but also decreases the stored amounts of potentially hazardous intermediates, including high-potency API. Increasing the use of on-line monitoring and control also reduces the burden of final testing, which mirrors the on-line control present in other continuous-manufacturing industries. Simulations of processes that include recycle loops demonstrated that improvements in process yield and robustness can be achieved by operating continuously [81, 82]. In spite of these promising results, there are still many hurdles to be overcome during the implementation of continuous processes. These include development of flow chemistry transformations, difficulties with processing dry solids and solid-laden fluids, lack of equipment at bench and pilot scale, development of control methodologies to guarantee product quality, and breaks in the process, especially between synthesis and formulation. Many examples have been reported of continuous processes for chemical synthesis in flow, reactions with workup, continuous crystallisation, drying, powder blending, and tableting. However, only a few others have considered multi-step portions of a process [83].

### 2.3.7 Regulatory Issues Related to Hot-Melt Extrusion

Pharmaceuticals continue to have an increasingly prominent role in health care. Therefore pharmaceutical manufacturing will need to employ innovation, cutting edge scientific and engineering knowledge, along with the best principles of quality management to respond to the challenges of new discoveries (e.g., novel drugs and nanotechnology) and ways of doing business (e.g., individualised therapy, genetically tailored treatment). Regulatory policies must also rise to the challenge.

### *2.3.8 Process Analytical Technology Framework*

Within the PAT framework, a process end point is not a fixed time; rather it is the achievement of the desired material attribute (**Figure 2.5**). This, however, does not mean that process time is not considered. A range of acceptable process times (process window) is likely to be achieved during the manufacturing phase and should be evaluated, and considerations for addressing significant deviations from acceptable process times should be developed [84]. This is one way to be consistent with relevant current good manufacturing practice (CGMP) requirements, as such control procedures that validate the performance of the manufacturing process [21 CFR 211.110(a)]. World and co-workers have classified the PAT applications into five levels as follows [85, 86]:

PAT Level 1: Determination of the concentration and other properties of component from the spectrum.

PAT Level 2: Raw material quality check and the classification from spectral data.

PAT Level 3: Batch process monitoring by means of multivariate measurement.

PAT Level 4: Using the batch monitoring data and process information to predict the quality of the product.

PAT Level 5: Feedback control of the process to maintain the quality of the product.

## 2.4 Conclusion

HME-enabled continuous pharmaceutical manufacturing needs a critical quality monitoring control system of essential process factors for every component of the batch. Proper generation of characterisation data, spectral analysis results and multivariate monitoring observations are useful for the understanding of the entire process. This can be accomplished by applying PCA to the observed spectral results. In this chapter, the role of in-line process monitoring techniques with the use of Raman, NIR for monitoring and regulating control of a continuous HME process is described with a predictive flow chart model. The chapter provides information regarding software for PAT, the applications of PAT and available marketed HME-based product. The controlling systems such as MPC, PCA, PLS have to be applied for continuous quality product development of drug delivery systems using HME. This chapter provides illustrative information which could be implemented for the design and development of continuous HME-based product.

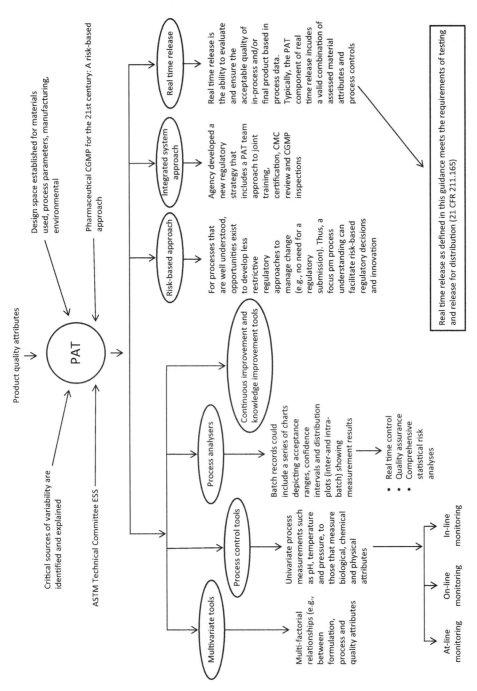

**Figure 2.5** Represents process parameter information in PAT approach. ASTM: American Society for Testing and Materials

## References

1.  D. Treffer, P. Wahl, D. Markl, G. Koscher, E. Roblegg and J.G. Khinast, *AAPS Advances in the Pharmaceutical Sciences Series*, 2013, **9**, 363.

2.  K. Plumb, *Chemical Engineering Research and Design*, 2005, **83**, 730.

3.  R. Singh, R. Rozada-Sanchez, T. Wrate, F. Muller, K.V. Gernaey, R. Gani and J.M. Woodley, *Computer Aided Chemical Engineering*, 2011, **29**, 291.

4.  E.H. Stitt, *Chemical Engineering Journal*, 2002, **90**, 47.

5.  T. Vasconcelos, B. Sarmento and P. Costa, *Drug Discovery Today*, 2007, **12**, 1068.

6.  R. Fule, T. Meer, A. Sav and P. Amin, *Journal of Pharmaceutical Investigation*, 2013, **43**, 305.

7.  C.V. Möltgen, T. Herdling and G. Reich, *European Journal of Pharmaceutics and Biopharmaceutics*, 2013, **85**, 1056.

8.  *Guidance for Industry: PAT-A Framework for Innovative Pharmaceutical Manufacturing & Quality Assurance*, US Food and Drug Administration, Silver Spring, MD, USA, 2004. *http://www.fda.gov*

9.  R. Fule, T. Meer, A. Sav and P. Amin, *Journal of Pharmaceutics*, 2013, **1**, 1.

10. M. Andersson, S. Folestad, J. Gottfries, M.O. Johansson, M. Josefson and K-G. Wahlund, *Analytical Chemistry*, 2000, **72**, 2099.

11. R. Marbach, *Journal of Near Infrared Spectroscopy*, 2005, **13**, 241.

12. R. Fule, T. Meer, P. Amin, D. Dhamecha and S. Ghadlinge, *Journal of Pharmaceutical Investigation*, 2014, **44**, 1, 41.

13. R. Fule and P. Amin, *Asian Journal of Pharmaceutical Sciences*, 2014, **9**, 2, 92.

14. H.S. Kaufman and J.J. Falcetta in *Introduction to Polymer Science and Technology: An SPE Textbook*, 1st Edition, John Wiley & Son, NY, USA, 1977.

15. J.P. Puaux, G. Bozga and A. Ainser, *Chemical Engineering Science*, 2000, **55**, 1641.

16. P.S. Johnson in *Developments in Extrusion Science and Technology*, Polysar Technical Publication No.72, Polysay Limited, South Sarnia, Ontario, Canada, 1982.

17. M.A. Repka, J.W. McGinity, F. Zhang and J.J. Koleng in *Encyclopedia of Pharmaceutical Technology*, Eds., J. Swarbrick and J. Boylan, Marcel Dekker, NY, USA, 2002a, **2**, 203.

18. K. Luker in *Pharmaceutical Extrusion Technology*, Eds., I. Ghebre-Sellassie and C. Martin, Marcel Dekker, NY, USA, 2003, **133**, 39.

19. M. Mollan in *Pharmaceutical Extrusion Technology*, Eds., I. Ghebre-Sellassie and C. Martin, Marcel Dekker, NY, USA, 2003, **133**, 1.

20. J. Breitenbach, *European Journal of Pharmaceutics and Biopharmaceutics*, 2002, **54**, 107.

21. W. Thiele in *Pharmaceutical Extrusion Technology*, Eds., I. Ghebre-Sellassie and C. Martin, Marcel Dekker, NY, USA, 2003, **133**, 69.

22. C. Rauwendaal in *Polymer Extrusion*, 3rd Edition, Hanser/Gardner Publications, Cincinnati, OH, USA, 1994.

23. T. Whelan and D. Dunning in *The Dynisco Extrusion Processor Handbook*, 1st Edition, London School of Polymer Technology, London, UK, 1996.

24. P.A. Hailey, P. Doherty, P. Tapsell, T. Oliver and P.K. Aldridge, *Journal of Pharmaceutical Biomedical Analysis*, 1996, **14**, 551.

25. R. Jayachandra Babu, W. Brostow, I.M. Kalogeras and S. Sathigari, *Material Letters*, 2009, **63**, 2666.

26. A.J. Kovacs, *Journal of Polymer Science*, 1958, **30**, 131.

27. W. Brostow, R. Chiu, I.M. Kalogeras and A. Vassilikou-Dova, *Materials Letters*, 2008, **62**, 3152.

28. V.P. Privalko, *Journal of Materials*, 1998, **20**, 57.

29. Hartung, M. Knoell, U. Schmidt and P. Langguth, *Drug Development and Industrial Pharmacy*, 2010, **37**, 3, 274.

30. R. Fule, and P. Amin, *BioMed Research International*, 2014, DOI.org/10.1155/2014/146781.org/10.1155/2014/146781.

31. D.S. Hausman, R.T. Cambron and A. Sakr, *International Journal of Pharmaceutics*, 2005, **298**, 80.

32. A. Heinz, C.J. Strachan, K.C. Gordon and T. Rades, *Journal of Pharmacy and Pharmacology*, 2009, **61**, 971.

33. D.C. Hinz, *Analytical Bio-analytical Chemistry*, 2006, **384**, 1036.

34. M.M. Crowley, F. Zhang, MA. Repka, S. Thumma, S.B. Upadhye, SK. Battu, J.W. McGinity and C. Martin, *Drug Development and Industrial Pharmacy*, 2007, **33**, 9, 909.

35. Y.R. Hu, H. Wikström, S.R. Byrn and L.S. Taylor, *Applied Spectroscopy*, 2006, **60**, 977.

36. J. Johansson and S. Folestad, *European Pharmaceutical Review*, 2003, **8**, 36.

37. A.C. Jorgensen, J. Rantanen, P. Luukkonen, S. Laine and J. Yliruusi, *Analytical Chemistry*, 2004, **76**, 5331.

38. L. Dierickx, L. Saerens, A. Almeida, T. De Beer, J.P. Remon and C. Vervaet, *European Journal of Pharmaceutics and Biopharmaceutics*, 2012, **81**, 683.

39. L. Dierickx, J.P. Remon and C. Vervaet, *European Journal of Pharmaceutics and Biopharmaceutics*, 2013, **85**, 1157.

40. A.C. Jorgensen, P. Luukkonen, J. Rantanen, T. Schaefer, A.M. Juppo and J. Yliruusi, *Journal of Pharmaceutical Sciences*, 2004, **93**, 2232.

41. *US Department of Health (2011) Guidance for Industry: 21 CFR Part 11; Electronic Records; Electronic Signatures, Glossary of Terms*, Volume 1, US Food and Drug Administration, Rockville, MD, USA, 2011.

42. K. Kamada, S. Yoshimura, M. Murata, H. Murata, H. Nagai, H. Ushio and K. Terada, *International Journal of Pharmaceutics*, 2009, **368**, 103.

43. *Guidance for Industry, Process Validation General Principles and Practices*, US Food and Drug Administration, Silver Spring, MD, USA, January, 2011, p.1.

44. J.D. Kirsch and J.K. Drennen, *Pharmaceutical Research*, 1996, **13**, 234.

45. K. Kogermann, J. Aaltonen, C.J. Strachan, K. Pollanen, J. Heinamaki, J. Yliruusi and J. Rantanen, *Journal of Pharmaceutical Sciences*, 2008, **97**, 4983.

46. T. De Beer, A. Burggraeve, M. Fonteyne, L. Saerens, J.P. Remon and C. Vervaet, *International Journal of Pharmaceutics*, 2011, **417**, 32.

47. G.M. Troup and, C. Georgakis, *Computers & Chemical Engineering*, 2013, **51**, 157.

48. S. Wartewig and R.H.H. Neubert, *Advanced Drug Delivery Reviews*, 2005, **57**, 1144.

49. M.J. Lee, C.R. Park, A.Y. Kim, B.S. Kwon, K.H. Bang, Y.S. Cho, M.Y. Jeong and G.J. Choi, *Journal of Pharmaceutical Sciences*, 2010, **99**, 325.

50. C. Gendrin and Y. Roggo and C. Collet, *Journal of Pharmaceutical and Biomedical Analysis*, 2008, **48**, 533.

51. M. Fransson and S. Folestad, *Chemometrics and Intelligent Laboratory Systems*, 2006, **84**, 1–2, 56.

52. K.M. Kleissas, A. Chong and B.T. Thompson, *American Pharmaceutical Review*, 2007, **10**, 4, 72.

53. A. Peinado, J. Hammond and A. Scott, *Journal of Pharmaceutical and Biomedical Analysis*, 2011, **54**, 1, 13.

54. L. Schenck and G.M. Troup in *Engineering in the Pharmaceutical Industry*, John Wiley & Sons, Inc., NY, USA, 2011, p.819.

55. L. Saerens, L. Dierickx, B. Lenain, C. Vervaet, J.P. Remon and T. De Beer, *European Journal of Pharmaceutics and Biopharmaceutics*, 2011, **77**, 1, 158.

56. S. Adam, D. Suzzi, C. Radeke and J.G. Khinast, *European Journal of Pharmaceutical Sciences*, 2011, **42**, 106.

57. S.C. Pinzaru, I. Pavel, N. Leopold and W. Kiefer, *Journal of Raman Spectroscopy*, 2004, **35**, 338.

58. W. Li, M.C. Johnson, R. Bruce, H. Rasmussen and G.D. Worosila, *Journal of Pharmaceutical and Biomedical Analysis*, 2007, **43**, 711.

59. M.V. Pellow-Jarman, P.J. Hendra and R.J. Lehnert, *Vibrational Spectroscopy*, 1996, **12**, 257.

60. P. Luukkonen, M. Fransson, I.N. Bjorn, J. Hautala, B. Lagerholm and S. Folesta, *Journal of Pharmaceutical Sciences*, 2008, **97**, 950.

61. S. Radl, T. Tritthart and J.G. Khinast, *Chemical Engineering Science*, 2010, **65**, 1976.

62. H.P. Zhu, Z.Y. Zhou, R.Y. Yang and A.B. Yu, *Chemical Engineering Science*, 2008, **63**, 5728.

63. Y. Roggo, P. Chalus, L. Maurer, C. Lema-Martinez, A. Edmond and N. Jent, *Journal of Pharmaceutical and Biomedical Analysis*, 2007, **44**, 3, 683.

64. J. Luypaert, S. Heuerding, Y. Vander Heyden and D.L. Massart, *Journal of Pharmaceutical Biomedical Analysis*, 2004, **36**, 495.

65. J. Märk, M. Karner, M. Andre, J. Rueland and C.W. Huck, *Analytical Chemistry*, 2010, **82**, 4209.

66. A. Lekhal, K. Girard, M. Brown, S. Kiang, B. Glasser and J.G. Khinast, *Powder Technology*, 2003, **132**, 119.

67. A. Lekhal, K.P. Girard, M.A. Brown, S. Kiang, J.G. Khinast and B.J. Glasser, *International Journal of Pharmaceutics*, 2004, **270**, 263.

68. R. Singh, M. Ierapetritou and R. Ramachandran, *European Journal of Pharmaceutics and Biopharmaceutics*, 2013, **85**, 1164.

69. J. Richalet, *Automatica*, 1993, 29, **5**, 1251.

70. C.E. Garcia, D.M. Prett and M. Morari, *Automatica*, 1989, **25**, 3, 335.

71. K.R. Muske and J.B. Rawlings, *AIChE Journal*, 1993, **39**, 262.

72. M.L. Darby and M. Nikolaou, *Control Engineering Practice*, 2012, **20**, 328.

73. D. Dougherty and D. Cooper, *Control Engineering Practice*, 2003, **11**, 141.

74. R. Singh, M. Ierapetritou and R. Ramachandran, *European Journal of Pharmaceutics and Biopharmaceutics*, 2013, **85**, 1164.

75. C.M. McGoverin, T. Rades and K.C. Gordon, *Journal of Pharmaceutical Sciences*, 2008, **97**, 4598.

76. J. Muller, K. Knop, J. Thies, C. Uerpmann and P. Kleinebudde, *Drug Development and Industrial Pharmacy*, 2010, **36**, 234.

77. Y. Li, H. Pang, Z. Guo, L. Lin, Y. Dong, G. Li, M. Lu and C. Wua, *Journal of Pharmacy and Pharmacology*, 2013, **66**, 148.

78. J. Rantanen, E. Rantanen, J. Tenhunen, M. Kansakoski, J.P. Mannermaa and J. Yliruusi, *European Journal of Pharmaceutics and Biopharmaceutics*, 2000, **50**, 271.

79. J. Rantanen, H. Wikström, R. Turner and L.S. Taylor, *Analytical Chemistry*, 2005, **77**, 556.

80. J. Rantanen, *Journal of Pharmacy and Pharmacology*, 2007, **59**, 171.

81. G. Reich, *Advanced Drug Delivery Reviews*, 2005, **57**, 1109.

82. A.S. Rathore, R. Bhambure and V. Ghare, *Analytical and Bioanalytical Chemistry*, 2010, **398**, 1, 137.

83. *US Department of Health: Guidance for Industry: PAT – A framework for Innovative Pharmaceutical Development, Manufacturing and Quality Assurance*, US Food and Drug Administration, Rockville, MD, USA, 2004.

84. S. Wold, J. Cheney, N. Kettaneh and C. McCready, *Chemometric and Intelligent Laboratory Systems*, 2006, **84**, 159.

85. L. Saerens, L. Dierickx, B. Lenain, C. Vervaet, J.P. Remon, and T. De Beer, *European Journal of Pharmaceutics and Biopharmaceutics*, 2010, **77**, 158.

86. S. Mascia, P.L. Heider, H. Zhang, R. Lakerveld, B. Benyahia, P.I. Barton, R.D. Braatz, C.L. Cooney, J.MB. Evans, T.F. Jamison, K.F. Jensen, AS. Myerson and B.L. Trout, *Angewandte Chemie*, 2013, **52**, 47, 12359.

# 3 Co-extrusion as a Novel Approach in Continuous Manufacturing Compliance

Mohammed Maniruzzaman

## 3 Background

This chapter appraises the role of hot-melt co-extrusion (Co-HME) in the pharmaceutical industry as an advanced technique. The hot-melt method enjoyed a renaissance with the implementation of hot-melt extrusion (HME) in the pharmaceutical industry [1]. HME has the advantage to be scalable and industrially applicable [2]. Duration of temperature elevation is reduced compared with traditional melting methods and hence processing of somewhat thermolabile active pharmaceutical ingredients (API) is enabled [3].

## 3.1 Introduction

To date, HME has been seen utilised significantly to develop and optimise solid dispersions (dispersions of two or more solids onto an inert matrix) in order to enhance the dissolution rate of water-insoluble drugs [4–7]. It has also been reported in literature that HME can successfully be implemented to mask the bitter taste of unpleasant drugs for the development of paediatric medicines as well as sustained or controlled-release (CR) drug delivery systems [5–8]. HME is referred in general as a process for forcing raw materials in a heated barrel with rotating screw(s) under elevated temperature into a product of uniform shape [8, 9]. In HME processing, mainly binary mixtures of powdered formulations (e.g., drug with polymers or lipids) are used. In contrast, Co-HME is defined as a process where melt extrusion is carried out simultaneously in the presence of two or more carriers (matrices) along with the active substance. This process aims at optimising and developing multi-layered extrudates at the end of the process. In most cases, in Co-HME processing, extrudates are optimised and collected as hot dry powders or agglomerated granules without the use of a die. The Co-HME process can also be coupled with novel approaches such as wet granulations or hot granulations [10]. Recently, a study has been reported describing a process *via* the HME technique to manufacture amorphous solid dispersions *via* a hot-melt processing method [10]. However, it is widely held that these reported works have primarily focused on the use of inorganic excipients alone as carriers in HME processing, and that none of them investigated the simultaneous use of inorganic excipients with co-processed materials such as polymer or plasticiser. There is an immense need to explore novel opportunities in this emerging subject area.

The simultaneous extrusion of graphite and presswood for making pencils has already been patented in the 19th century [10]. Since 1940, the Co-HME was utilised predominantly in the plastics industry and to a lesser extent in the food industry. Co-extrusion of plastics started with the production of pipes, wires and cables. An early example of plastic co-extrusion is the multilayer drinking straw that came on the market in 1963. Around 1984, co-extrusion became popular in the food industry to produce snacks with different colours, textures, or flavours.

The pharmaceutical industry became interested in this innovative technology. It is an innovative continuous production technology that offers numerous advantages over traditional pharmaceutical processing techniques. The technique allows for the combining of the desirable properties of multiple materials into a single structure with enhanced performance characteristics. This makes Co-HME an excellent alternative to other conventionally available techniques such as solvent evaporation, freeze-drying, spray drying and so on. Besides the continuity of the process, its major advantages are fewer processing steps, no use of organic solvents/water and the possibility of improving the solubility or for sustaining drug release. This technique is full of promises but so far, there are no co-extruded dosage forms for oral use on the market and only a few papers have been published during the last decades.

## 3.2 Applications of Co-extrusion *via* Hot-Melt Extrusion in Continuous Manufacturing

HME is the process of melting, plasticising and mixing a blend of drug and carrier inside a heated barrel. The molten/plasticised material is transferred through the barrel using one or two rotating screws and subsequently pressed through a die into a product of uniform shape and high density [11]. This is a most advanced and widely used manufacturing process which has been used in the plastic industry since the 1930s. Over the last two decades, the pharmaceutical industry and academia have gradually focused on the potential of HME as a continuous production process in the pharmaceutical industry [12]. HME has been investigated in a broad range of pharmaceutical applications for drug delivery *via* oral and transdermal routes [13, 14]. While HME has proven to be a successful processing technique, used in pharmaceutical industry to produce drug products in a continuous way, co-extrusion is quite new in pharmaceutical applications [15, 16]. Nevertheless, co-extrusion of polymers is widely applied in the plastics and packaging industries. The pharmaceutical co-extrusion process consists of the simultaneous HME of two or more drug loaded formulations creating a multi-layered extrudate. HME as a continuous manufacturing technology has shown some other major advantages over conventional techniques, such as improving the bioavailability of poorly water-soluble drugs *via* molecular dispersions [17], without the requirement for processes based on organic solvent or aqueous spray drying. Moreover, *via* HME, matrix formulations can be manufactured using polymers that act as drug depots [14]. The added value of co-extrusion is that it the release of each drug can be modulated independently, to enable simultaneous administration of non-compatible drugs and to produce fixed-dose combinations in a continuous single-step process. By processing the co-extrudate into mini-matrices that can be easily filled into gelatin capsules, a multi-particulate formulation is created. A specific challenge during co-extrusion is to establish a core/coat polymer combination fit for purpose which takes into account the required release characteristics of the incorporated drugs, similarity in extrusion temperature and appropriate adhesion between the layers. So far, no co-extruded dosage forms for oral use are on the market.

### 3.2.1 Problem Encountered During Co-extrusion

The development of a co-extruded formulation is challenging as technical considerations have to be taken into account when selecting polymer combinations.

### 3.2.2 Variation of the Die Temperature

The first challenge is to extrude both layers through one die, under the same temperature setting of the die. Although both melts may be extruded at different temperature conditions in each barrel, each temperature profile must enable the flow of the melts through the die at the set temperature [18]. In order to do a successful co-extrusion, the melts should be processed at a similar temperature [15].

### 3.2.3 Swelling of the Die

The cross section of extrudates can increase upon leaving the die. This phenomenon is described as 'die swell' and depends on external factors and the viscoelastic properties of the polymers [15]. Die swell needs extra attention in co-extrusion since it can occur in each layer and thereby influence the adhesion between both layers.

### 3.2.4 Viscosity Matching

Layer non-uniformity may be a problem during co-extrusion and can be caused by many process factors such as velocity mismatch. The key for success is the adequate choice of materials in order to match the melt viscosities of the layers [19]. Polymer viscosity depends on shear and temperature. Viscosity matching is not always straightforward since each polymer has its own viscoelastic properties and each layer is exposed to different shear rates during processing. Viscosity mismatch can cause encapsulation when a low viscosity polymer flows around a high viscosity compound and encapsulates it. Viscosity mismatch may also yield interfacial instabilities, such as zigzag and wave patterns. The multi-manifold dies allow for a larger mismatch in viscosity, since the melts are combined near the end of the die.

### 3.2.5 Adhesion

Adequate adhesion is essential to avoid separation during downstream processing [19]. Adhesion is defined as the tendency of dissimilar particles or surfaces to cling to one another or, as the molecular attraction that holds the surfaces of two dissimilar substances together [19]. There are 3 main mechanisms of adhesion: 1) mechanical interlocking; 2) molecular bonding; and 3) thermodynamic adhesion [20]. 'Mechanical interlocking adhesion' occurs if the material of one layer interlocks into the irregularities of the surface of the other layer [19]. 'Molecular adhesion' can occur when surface atoms of two separate surfaces can interact. These interactions include dipole-dipole interactions, van der Waals forces and chemical interactions. 'Thermodynamic adhesion' implies that the thermodynamics of the polymer system will try to minimise the surface free energy [21].

### 3.2.6 Interdiffusion

Miscible polymers will diffuse into each other at the interface and may in this way facilitate adhesion, whereas immiscible polymers generally exhibit weaker adhesion due to limited diffusion. Although polymer compatibility influences adhesion, it is not a prerequisite. According to Dierickx and co-workers [22] adhesion may also be influenced by many process variables including temperature, contact time and drug load. In certain cases, it may be appropriate to study the degree of interdiffusion of the layers. If adjacent layers are miscible, the API and/or polymer of one layer may diffuse into the other layer. Innovative spectral methods (Raman mapping) have been used to study migration at the core/coat interface [23].

### *3.2.7 Delamination*

Preventing delamination of co-extrudates is another challenge. Delamination can occur when the shrinkage percentages of the polymers differ too much. The role of shrinking may become clear if one imagines a co-extruded tube of two concentric layers. If the inner layer shrinks more than the outer layer, delamination can occur since a gap will arise between both layers. If the outer layer is brittle and shrinks more than the inner layer it might burst or on the other hand, if the outer layer is stretchable, it might tighten around the inner layers and a delamination free structure may be achieved [24].

## 3.3 Pharmaceutical Applications

The birth control thermoplastic device known as Nuvaring releases etonogestrel and ethinyl estradiol and needs to be left in the vagina for 3 weeks. Mainly it contains two steroid drugs molecularly dispersed within polyethylene vinyl acetate copolymers and is enveloped with a thin polymer membrane. Polyethylene vinyl acetate polymers with a lower vinyl acetate fraction serve as a rate-limiting membrane to control drug release [25–27]. Implanon® (Schering-Plough), a non-biodegradable, flexible and implantable rod that contains etonogestrel, provides contraceptive efficacy for 3 years [28]. The rod consists of a solid core, etonogestrel crystals embedded in ethylene vinyl acetate, and is surrounded by an outer ethylene vinyl acetate membrane that controls the release rate. The polymers have a different vinyl acetate content in the outer and inner layers [29].

### *3.3.1 Pharmaceutical Significance of the Co-extrusion Technique*

So far, there is no co-extruded drug on the market. There are only few papers which have been published on co-extrusion for oral drug delivery systems [16, 19, 30–32]. Day-by-day, it's increasingly gaining importance for oral drug delivery system [15, 17]. Combination therapy has been recognised as a major advantage, not only for patent elongation but also for therapeutic reasons [33, 34]. The combination of active substances within a single pharmaceutical form of administration is a so-called fixed-dose combination or fixed combination medicinal product. According to the European Medicine Agency [35], fixed-dose combination or fixed combination medicinal products have been increasingly used to improve compliance by simplification of the therapy, or to benefit from the added effects of two medicinal products given together. Other advantages may include the counteracting of adverse reactions [35, 36].

Co-extrusion enables a wide range of formulation strategies for oral dosage forms as it offers the opportunity to optimise drug release of an individual layer [25]. Hence, dosage forms may be developed wherein one layer (e.g., the core) exhibits CR and another layer (e.g., the coat), an immediate release property [8]. Another approach might be to combine two immediate release layers. These 'technical' opportunities combined with the repertoire of available polymers permit the modulation of the drug release rate of a formulation. Secondly, co-extrusion allows the final drug product to contain one, two or more API. A dual release of one API was achieved by incorporating the API in an immediate release layer in the coat, and in a CR layer in the core [19, 22]. Obviously, also different (incompatible) API can be incorporated into different layers, each with their desired release properties [15, 36]. The unique properties of an API together with the desired release profile may provide a rationale for polymer selection in a certain layer.

Successful formulations with an immediate release layer in the coat and a CR in the core have already been developed [15, 19, 22, 36]. However, a lot of undiscovered opportunities remain as the combination of two or more immediate release layers has not yet been explored. In addition, co-extrusion enables the manufacturing of a combination product with a highly soluble and poorly soluble drug in two separate layers.

## 3.4 Case Study

The case study involves the use of an amorphous magnesium aluminometasilicate (MAS) as a co-processing excipient with cationic methacrylate hydrophilic polymer (MEPOL) for the formation of a solid dispersion of drug A. Drug A is a crystalline drug with poor water-solubility. The solid-state characteristics of the API are reported together with details of a possible intermolecular interaction. Drug A formulations with MAS/MEPOL (20–40% drug w/w) were well mixed in 100 g batches for 10 min each prior to extrusion. A Turbula TF2 Mixer was used to blend the powder formulations thoroughly. Extrusion of all formulations was performed using a Eurolab 16 mm twin-screw extruder (Thermo Fisher, Germany) equipped with a 2 mm rod die with a screw speed of 50 rpm (feed-rate 0.6 kg/h). The temperature profile used for all formulations was 50/100/120/135/135/135/135/135/135/135 °C (from feeding zone → die). The produced extrudates (strands) were cut into 1 mm pellets by using a pelletiser (Thermo Fisher, Germany). The pellets were micronised by a rotor milling system to collect the granules below 250 μm threshold (Retsch, Germany). The extrudates were characterised *via* differential scanning calorimetry (DSC), X-ray powder diffraction (XRPD), scanning electron microscopy (SEM), Fourier-Transform infrared (FT-IR) and *in vitro* dissolution studies.

The SEM results revealed no Drug A crystals on the surface of the extrudates. DSC thermograms of the physical mixtures of the Drug A/MAS/MEPOL (20–40%) showed two endothermic thermal transitions corresponding to the glass transition temperature ($T_g$) of the polymer (~59 $^{o}$C) and the melting peak of the drug at relatively higher temperature (166–187 $^{o}$C). None of these transitions are detected in extruded formulations apart from a $T_g$ shifted towards a lower temperature in all extrudates indicating the presence of the molecularly dispersed Drug A ($T_g = 42$ $^{o}$C) in the MAS/MEPOL matrices. The XRPD analysis showed no distinct peaks representing the crystalline state in any of the extruded formulations, while Drug A occurred in the crystalline form in all physical mixtures.

Drug A exists as dimers in the crystalline state due to the benzoyl, carbonyl and carboxylic acid containing structure. The FT-IR spectra of the crystalline state of Drug A showed the acid dimer peak at 1,715 cm$^{-1}$ and at 1,690 cm$^{-1}$, respectively. The absorbance of the drug dimer peaks start getting broader and shift to 1,745 cm$^{-1}$ with the increase of Drug A loadings, indicating the amorphicity of Drug A in the extruded formulations. Furthermore, a new absorbance peak at 2,950 cm$^{-1}$ in extrudates indicates a possible intermolecular interaction between the MEPOL and the drug. Further investigations using X-ray photoelectron spectroscopy and/or energy dispersive X-ray could provide further evidence of such interactions. As expected, the extrudates with 20–30% Drug A loadings showed faster release of Drug A (80% in 180 min at pH 6.8) compared to that of the physical mixtures at all drug/polymer ratios and the pure Drug A itself.

## 3.5 Conclusions

The co-HME technique shows great potential and has gained interest over the traditional extrusion technique. There are challenges in selecting the appropriate polymer combinations and in the design of the die which plays an important role for a successful extrusion technique. Continued research will help to further the benefits of the co-HME technique, especially as the ability to combine drug products will be of interest to the pharmaceutical industry.

## References

1.  C. Leuner and J. Dressman, *European Journal of Pharmaceutics and Biopharmaceutics*, 2000, **50**, 47.

2.  S. Janssens and G. van den Mooter, *Journal of Pharmacy and Pharmacology*, 2009, **61**, 1571.

3.  M.A. Repka, T.G. Gerding, S.L. Repka and J.W. McGinity, *Drug Development and Industrial Pharmacy*, 1999, **25**, 625.

4.  M. Maniruzzaman, J.S. Boateng, M. Bonnefille, A. Aranyos, J.C. Mitchell and D. Douroumis, *European Journal of Pharmaceutics and Biopharmaceutics*, 2012, **80**, 2, 433.

5.  M. Maniruzzaman, J.S. Boateng, M.J. Snowden and D. Douroumis, *International Scholarly Research Notices: Pharmaceutics*, 2012, DOI. org/10.5402/2012/436763.

6.  M. Maniruzzaman, M. Bonnefille, A. Arayonos, M.J. Snowden and D. Douroumis, *Journal of Pharmacy and Pharmacology*, 2014, **66**, 2, 323.

7.  M. Maniruzzaman, D.J. Morgan, A.P. Mendham, J. Pang, M.J. Snowden and D. Douroumis, *International journal of Pharmaceutics*, 2013, **443**, 1–2, 199.

8.  M. Maniruzzaman in *Development of Hot-Melt Extrusion as a Novel Technique for the Formulation of Oral Solid Dosage Forms*, Greenwich Academic Literature Archive, University of Greenwich, London, UK, 2012. [PhD Thesis]

9.  *Neusilin®*: The Specialty Excipient, Fuji Chemical Industry Co., Ltd, Toyama-Pref, Japan.

10. M.K. Gupta, Y-C. Tseng, D. Goldman and R.H. Bogner, *Pharmaceutical Research*, 2002, **19**, 11, 1663.

11. M.A. Repka, S.K. Battu, S.B. Upadhye, S. Thumma, M.M. Crowley, F. Zhang, C. Martin and J.W. McGinity, *Drug Development and Industrial Pharmacy*, 2007, **33**, 1043.

12. M.A. Repka, S. Majumdar, S.K. Battu, R. Srirangam and S.B. Upadhye, *Expert Opinion on Drug Delivery*, 2008, **5**, 1357.

13. M. Maniruzzaman, S.J. Boateng, B.Z. Chowdhry and D. Douroumis, *Drug Development and Industrial Pharmacy*, 2013, **39**, 2, 218.

14. M.M. Crowley, F. Zhang, M.A. Repka, S. Thumma, S.B. Upadhye, S.K. Battu, J.W. McGinity and C. Martin, *Drug Development and Industrial Pharmacy*, 2007, **33**, 909.

15. L. Dierickx, L. Saerens, A. Almeida, T. De Beer, J.P. Remon and C. Vervaet, *European Journal of Pharmaceutics and Biopharmaceutics*, 2012, **81**, 683.

16. U. Quintavalle, D. Voinovich, B. Perissutti, E. Serdoz, G. Grassi, A. Dal Col and M. Grassi, *European Journal of Pharmaceutical Sciences*, 2008, **33**, 282.

17. J. Breitenbach, M. Magerlein, I. Ghebre-Sellassie and C. Martin, *Pharmaceutical Extrusion Technology*, 2003, **133**, 245.

18. L. Dierickx, J.P. Remon and C. Vervaet, *European Journal of Pharmaceutics and Biopharmaceutics*, 2013, **85**, 1157.

19. W.W. Gerberich and M.J. Cordill, *Reports on Progress in Physics*, 2006, **69**, 2157.

20. F. Awaja, M. Gilbert, G. Kelly, B. Fox and P.J. Pigram, *Progress in Polymer Science*, 2009, **34**, 948.

21. Y.S. Lipatov, *Science and Engineering of Composite Materials*, 1995, 4, **35**.

22. L. Dierickx, B. Van Snick, T. Monteyne, T. De Beer, J.P. Remon and C. Vervaet, *European Journal of Pharmaceutics and Biopharmaceutics*, 2014, **88**, 2, 502.

23. K.C. Gordon and C.M. McGoverin, *International Journal of Pharmaceutics*, 2011, **417**, 151.

24. D.F. Li, T.S. Chung and W. Rong, *Journal of Membrane Science*, 2004, **243**, 155.

25. M.A. Repka, S. Shah, J. Lu, S. Maddineni, J. Morott, K. Patwardhan and N.N. Mohammed, *Expert Opinion on Drug Delivery*, 2012, **9**, 105.

26. J.C. DiNunzio, C. Brough, J.R. Hughey, D.A. Miller, R.O. Williams and J.W. McGinity, *European Journal of Pharmaceutics and Biopharmaceutics*, 2010, **74**, 340.

27. J.A.H. van Laarhoven, M.A.B. Kruft and H. Vromans, *International Journal of Pharmaceutics*, 2002, 232, 163.

28. J. Huber, *Contraception*, 1998, **58**, 85.

29. M.A. Fischer, *Journal of Obstetric Gynecologic and Neonatal Nursing*, 2008, **37**, 361.

30. C.M. Vaz, P.F.N.M. van Doeveren, R.L. Reis and A.M. Cunha, *Polymer*, 2003, **44**, 5983.

31. U. Quintavalle, D. Voinovich, B. Perissutti, F. Serdoz and M. Grassi, *Journal of Drug Delivery Science and Technology*, 2007, **17**, 415.

32. T. Iosio, D. Voinovich, M. Grassi, J.F. Pinto, B. Perissutti, M. Zacchigna, U. Quintavalle and F. Serdoz, *European Journal of Pharmaceutics and Biopharmaceutics*, 2007, **69**, 686.

33. P.S. Hiremath, S.A. Bhonsle, S. Thumma and V. Vemulapalli, *Recent Patents on Oral Combination Drug Delivery and Formulations*, 2011, **5**, 52.

34. J. Woodcock, J.P. Griffin and R.E. Behrman, *New England Journal of Medicine*, 2011, **364**, 985.

35. *Committee for Medicinal Products for Human Use: Guideline on Clinical Development of Fixed Combination Medicinal Products*, European Medicines Agency, London, UK, 2009, CHMP/EWP/240/95 Rev. 1.

36. A-K. Vynckier, L. Dierickx, J. Voorspoels, Y. Gonnissen, C. Vervaet and J. Remon in *Proceedings of the AAPS Annual Meeting and Exposition*, 14–18th October, Chicago, IL, US, 2012.

# 4  Solid-state Engineering of Drugs using Melt Extrusion in Continuous Process

Mohammed Maniruzzaman

## 4 Introduction

The recent advent of high throughput screening seems to have been responsible for a tremendous increase in the number of water resistant drug molecules failing to make it out of the discovery pipeline. This leaves pharmaceutical scientists with a key challenge to improve the aqueous solubility of these poorly water-soluble drugs. Amorphous forms of crystalline drug candidates (poorly water-soluble) have always been considered an emerging phenomenon in pharmaceutical sciences [1]. Unlike its crystalline counterpart, the amorphous form of a compound appears in a more highly energetic state which is advantageous in terms of increasing solubility and thus bioavailability. Various physical approaches including hot-melt extrusion (HME) have been reported to improve drug solubility which has often led to the formation of amorphous forms of crystalline drugs [1]. In most of the cases, either polymeric or lipidic carriers have been utilised. A further investigation on the use of various novel inorganic excipients to enhance dissolution rates of these poorly water-soluble drugs can provide a new insight into pharmaceutical research and product development.

## 4.1 Dissolution Enhancement

There is a clear market need to develop methods to make drugs more soluble; it has been reported in the region of 30% of pipeline drugs are poorly soluble influencing the cost and time of drug development. Similarly, large numbers of drugs have stability and compressibility issues, solutions to which are currently being explored by different crystal and particle engineering techniques.

### 4.1.1 Use of Novel Inorganic Excipients

Bahl and co-workers reported the effects of Neusilin as a novel inorganic excipient for increasing the amorphicity of indomethacin (BCS class II drug) (INM) and thus enhancing its dissolution rate [2]. The authors found that the lower the ratio of indomethacin to Neusilin US2, the faster was the amorphisation. Also higher humidity seemed to facilitate amorphisation more effectively at the lower ratio of indomethacin to Neusilin. The authors concluded their investigation by claiming that hydrogen bonding and surface interaction between metal ions of Neusilin and indomethacin inhibited the increase in the amorphicity of the drug which could have potential effects to increase its solubility. Later on the same group reported a follow up study evaluating the solubility and dissolution profiles of indomethacin (both crystalline and amorphous state) in the presence of Neusilin as an inorganic carrier [3]. They reported the effect of the ratio of indomethacin: Neusilin along with other factors (e.g., percent crystallinity of the resulting formulations and pH) on the dissolution rate of the processed amorphous indomethacin. The presence of both silicic acid and ions ($Mg^{2+}$ and $Al^{3+}$) from Neusilin in the dissolution media were found to cause the increase in the concentration of indomethacin and enhance its *in vitro* dissolution rates [3].

Vercruysse and co-workers reported a screening of theophylline tablets manufactured *via* twin-screw wet granulation in order to improve process understanding and knowledge of process variables that determine granules and tablets quality [4]. The study was undertaken involving a premix of theophylline anhydrate with carriers, followed by granulation with demineralised water. The results showed that the quality of granules and tablets could be optimised by adjusting specific process variables using a continuous twin-screw granulator in HME. Based on quite a similar approach, Maclean and co-workers later on, proposed a scalable process to make Sulindac-Neusilin, an amorphous drug complex using HME [5]. The dissolution properties of the resulting HME material was improved while maintaining similar physical stability as those manufactured in ball milling. The HME material was used to make tablets that were found to have better dissolution properties than tablets made from crystalline Sulindac. The authors claimed that the inorganic silicates such as Neusilin had, offered a better choice than organic polymers to stabilise the amorphous phase.

However, recently a newly developed concept of an HME based processing technique (a process where intense mixing is performed in the presence of an absorbent instead of water- *via* continuous HME processing), has been seen reported. In this approach, various novel excipients can be used with or without polymers. In this optimised process, novel excipient-based carriers can be used to manufacture homogenous one phase amorphous extrudates either alone or in the presence of hydrophilic polymers (**Figure 4.1**).

**Figure 4.1** Scanning electron microscopy (SEM) image of drug/polymer extrudates

In another separate study, hot-melt granulation was used to prepare ternary solid dispersion granules in which the drug was dispersed in a carrier and coated onto an adsorbent. Seven drugs including four carboxylic acid-containing drugs (BAY 12-9566, naproxen, ketoprofen and indomethacin), a hydroxyl-containing drug (testosterone), an amide-containing drug (phenacetin), and a drug with no proton-donating group (progesterone) were studied with lipidic carrier Gelucire 50/13 and hydrophilic carrier polyethylene glycol 8000 being used as dispersion carriers. Neusilin US2 (magnesium aluminosilicate) was used as the surface adsorbent [6]. In this study, two competing mechanisms were proposed to explain the complex changes observed in drug dissolution upon storage of solid dispersion granules. Conversion of the crystalline drug to the amorphous hydrogen bonded state seemed to increase dissolution. The solubility of the drug in Gelucire proved to be a very crucial factor in determining the predominant mechanism by governing the flux toward the surface of Neusilin. The study was concluded with an observation that the mobility for this phenomenon was provided by the existence of the eutectic mixture in the molten liquid state during storage [6].

Recently, in another study, a report describing the use of excipients-polymer [Eudragit® EPO (EPO)] blends for the formation of solid dispersions, characterises the solid-state of the active pharmaceutical ingredient(s) (API) alongside a possible intermolecular interaction [7]. In this study, a manufactured solid dispersion was incorporated into orally disintegrating tablet(s) (ODT) which showed faster disintegration. The results obtained from the aforementioned study showed that the drug revealed no crystals on the surface of the extrudates. Differential scanning calorimetry (DSC) thermograms of the physical mixtures of the extruded formulations 20–40% showed two endothermic thermal transitions corresponding to the glass transition temperature ($T_g$) of the polymer (~59 °C) and the melting peak of the drug at a relatively higher temperature (166–187 °C). None of these peaks were visible in extruded formulations apart from a $T_g$ shifted towards a lower temperature in all extrudates, indicating the molecular dispersion of the drug onto the matrices. Similar results had been observed in the X-ray powder diffraction analysis. The absorbance of the drug dimer peaks started getting broader and shifted to 1,745 cm$^{-1}$ with the increase of the drug loadings, indicating possible intermolecular hydrogen bonding interactions between the drug and the polymer. Absorbance at 2,950 cm$^{-1}$ in extrudates also indicated possible interactions *via* the amide group of the drug.

The molecular dispersion was expected to provide increased dissolution rates compared to those of the physical blends and the pure substance itself. As expected, the release profiles of three different extruded formulations were identical in terms of increasing the *in vitro* dissolution rate of the drug (**Figure 4.2**). In addition it was observed that increased drug loading (40%), provided slightly faster release rates in contrast with the 20–30% loadings, suggesting that the solubility of the drug had been increased [7].

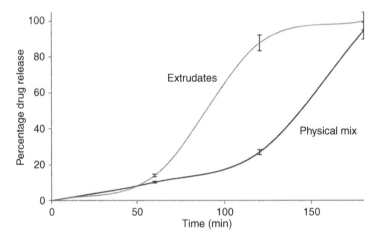

**Figure 4.2** *In vitro* drug release profiles in extruded formulations
(n = 3, paddle speed 100 revolutions per min at 37 ± 0.5 °C)

### 4.1.2 Use of Polymers to Enhance Dissolutions of Poorly Water-soluble Active Pharmaceutical Ingredients

Gryczke and co-workers processed up to 40% ibuprofen (IBU) – a poorly water-soluble drug, with EPO (50%) and talc (10%) and demonstrated a tremendous enhancement of the dissolution rate of IBU during the extrusion processing [8]. Results showed that increased IBU concentration was successfully used to facilitate strong drug–polymer interactions. IBU had also been found to show plasticising effects equated to those of traditional plasticisers. The presence of a single $T_g$ and the absence of IBU melting endotherm during the thermal analysis confirmed the complete miscibility of IBU/EPO and the creation of a glassy solution. IBU was found to be molecularly dispersed within EPO matrices, thus facilitating the higher dissolution rate and improving the taste masking efficiency. Subsequently, the ground extruded materials were compressed into tablets and compared with commercially available Nurofen® Meltlets Lemon ODT. Results showed that the extruded tablets were found to have about 5-fold increases in *in vitro* dissolution studies of poorly water-soluble IBU compared to that of Nurofen® (**Figure 4.3**) [8].

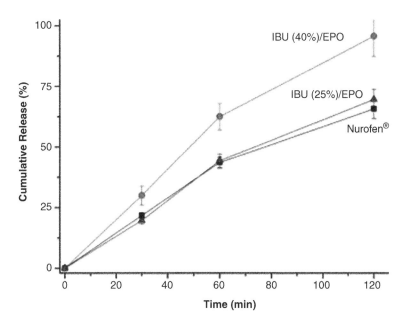

**Figure 4.3** Release profiles of ODT with IBU/EPO extruded granules of, (▲) 25%, (●) 40% drug loading and (■) Nurofen®. Adapted with permission from A. Gryczke, S. Schminke, M. Maniruzzaman, J. Beck and D. Douroumis, *Colloids and Surfaces B: Biointerfaces*, 2011, **86**, 2, 275 [8]

Similarly, Maniruzzaman and co-workers investigated the efficiency of two different hydrophilic polymeric carriers to enhance the dissolution rate of two poorly water-soluble API processed *via* HME [9]. INM and Famotidine (BCS IV) (FMT) were selected as model active substances while polyvinyl caprolactam graft copolymer, Soluplus® and vinyl pyrrolidone-vinyl acetate copolymer grades, Kollidon® VA 64 and Plasdone® S630 were used as hydrophilic polymeric carriers. The physicochemical properties and the morphology of the extrudates, evaluated *via* X-ray diffraction, DSC and SEM revealed the existence of amorphous drugs in all extruded formulations with all polymers. Increased drug loadings of up to 40% were achieved in the extruded formulations for both drugs where in the case of all polymers, molecular dispersions of the drug molecules were obtained. INM and FMT were seen to exhibit strong plasticisation effects with increasing drug concentrations. The *in vitro* dissolution studies showed increased INM/FMT release rates for all formulations compared to that of pure API alone, suggesting that HME had successfully been used to enhance the dissolution rate of various water resistant drugs.

### 4.1.3 Continuous Co-crystallisation Engineering

Many new drugs have poor physicochemical and biopharmaceutical properties which are considered real challenges in terms of constraining bioavailability, processing and clinical performance. The reduction in the numbers of new molecules coming to market along with the expiry of patents is a major concern for the pharmaceutical industry. This often results in repositioning and reformulation of the drugs. There is a clear market need to develop methods to make drugs more soluble. It has been reported that about 40% of the drugs in the discovery pipeline are poorly soluble or insoluble thereby increasing the cost and time of drug development [10]. Similarly, large numbers of drugs have stability and compressibility issues, solutions which are currently being explored by crystal and particle engineering techniques. There are various potential methods to overcome issues such as solubility. These include the use of particle engineering to reduce the size of the drug particles and forming salts or solid solutions by dissolving the drug in a soluble polymer. These techniques are generally complex and are only suitable for some types of drugs. Another potential method of improving the solubility of certain drugs is to form a co-crystal of the drug with another pharmaceutically accepted material, such as a sugar.

Pharmaceutical co-crystals are crystals which contain two or more neutral components (both solid at ambient temperatures), present in stoichiometric amounts. To date, co-crystals are an emerging interest in pharmaceutical drug development to improve the solubility, dissolution and thus bioavailability [11–14] of various poorly water-soluble drugs. Co-crystals enable the modification of key physicochemical properties of pharmaceuticals (e.g., stability) that impact on processing, pharmacokinetics, efficacy, toxicity, stability and design of the final dosage forms. As a result, there is growing interest in the development of co-crystals of API. Co-crystals are a modified/engineered form of the API made from the drug itself with host components, which are known as co-formers. Co-crystals have been reported by various sources as 'supramolecular crystals' in which molecules are held together by low energy non covalent interactions such as hydrogen bonding, van der Waals interactions or $\pi$-$\pi$ interactions [11, 15]. Host co-formers are normally non-covalently attached to the API without changing the chemical and physiological action of drug compound.

The 'solvent growth method' and 'mechanical method' are two most common techniques that have been used to make pharmaceutical co-crystals [16]. But in reality none of the techniques except the solvent growth method is scalable and all of them are time consuming. Excessive use of solvent can be harmful and costly as well and small residues of solvent can be toxic which can again raise regulatory issues. Another disadvantage in using the solvent growth method is the dispersion of two molecules in same solvent, which is not always possible as it creates equilibrium solubility difficulties. Some other techniques such as spray drying [16], ultrasonication [17], and supercritical fluid technique [11, 18] have also been used for the development and engineering of various pharmaceutical co-crystals [19].

Recently, HME has been reported as being successfully utilised for the manufacture of pharmaceutical co-crystals [20–24]. HME can either be processed with single-screw or twin-screw instruments. Generally, a single-screw extruder has a single-screw with conveying properties built as close to barrel as possible to produce sufficient shear. In contrast, a twin-screw extruder can be adapted with various screw designs and configurations as required by the formulations and the final dosage forms. The screw geometry of the twin-screw (co-rotating or counter-rotating) produces the highest shear and offers excellent mixing capacity in the barrel. The main advantage of HME techniques includes its viability for continuous processing without the use of solvents and it is also an economically friendly and easy to scale-up process [8, 25, 26].

### 4.1.4 Polymorphic Transformations via Hot-Melt Extrusion

In various literatures, it has been demonstrated that the molecular nature of solid-state dispersions and/or phase separation in melt extruded formulations can lead to crystallisation or polymorphic conversions of the API during the extrusion process. However, identification, prediction and characterisation of solid-state dispersions involving polymorphic conversions is difficult. The relationship between solid-state dispersions and their suitability as drug delivery systems is neither well understood nor predictable.

## 4.1.4.1 Polymorphic Transformation of Carbamazepine

A polymorphism is an important consideration during the pharmaceutical manufacturing process. Especially a metastable polymorph may accelerate physicochemical characterisation compared to a marketed form. In addition, the choice of carrier or excipient makes a significant impact on the solubility and dissolution of the active entity. In a research case study, a hydrophilic carrier, D-gluconolactone was extruded with model drug carbamazepine (CBZ) in various molecular ratios by using twin-screw hot-melt extrusion. The extruded mixtures were successfully analysed by powder X-ray diffraction, DSC, hot stage microscopy (HSM) and dissolution studies for drug release. All hot-melt extruder processed samples exhibited a polymorphic transformation to metastable from I and was propositional to the amount of CBZ. HME processed samples disclosed remarkable dissolution rates compared to pure CBZ.

Fewer changes were noted in dissolution within the physical mixture and extruded binaries with the highest amount of CBZ compared to the rest. The HME technique was successfully applied to increase the dissolution of poorly water-soluble drug CBZ by using D-gluconolactone as a hydrophilic carrier. A 1.5:1 molar ratio between the drug and the carrier was determined as the limited ratio which achieved about 91% of drug release (about 40% more than pure API). Unexpected but beneficial polymorphic transformations were productively analysed with different analytical techniques. Conversion to triclinic form I and fraction amount of D-gluconolactone in 1.5:1 molar ratio played an important role in increasing the dissolution rate of the formulations. In addition, a single-step, scalable, and efficient HME technique was effectively utilised. In future, it will be also interesting to see the relation between enthalpy and drug ratio used to construct phase diagrams to distinguish the solid-solid interaction between the API and the carrier. Mechanical properties will be evaluated in terms of flow and compactability of powder.

## 4.1.4.2 Polymorphic Transformation of Artemisinin

The application of high temperature extrusion (HTE) to produce solvent-free metastable drug forms has been explored by Paradkar and his group (2012) [27]. The example presented is that of the drug artemisinin (art), which exhibits two polymorphic forms: orthorhombic and triclinic. In general, the commercially available orthorhombic form is considered thermodynamically stable and has lower water-solubility. In contrast, the metastable triclinic form has a higher dissolution rate. Recently, this same group has reported the generation of co-crystals using the HME process in which the API and the co-former are processed in a twin-screw extruder at the eutectic point of the mixture as well.

A novel, green, and continuous method for solid-state polymorphic transformation of artemisinin by HTE has recently been demonstrated. This communication described attempts to understand the mechanisms causing phase transformation during the extrusion process. Polymorphic transformation was investigated using mainly the thermal analysis performed by HSM and a model shear cell. In the reported study, it was claimed that, at high temperature, phase transformation from orthorhombic to the triclinic crystals was observed through a vapour phase. Under certain mechanical stress, the crystalline structure was heavily disrupted and the process was continuous. This exposed the new surfaces and accelerated the transformation process [28].

## References

1.  M. Maniruzzaman, J.S. Boateng, M.J. Snowden and D. Douroumis, *International Scholarly Research Notices: Pharmaceutics*, 2012, Article ID:436763.

2.  D. Bahl and R.H. Bogner, *Pharmaceutical Research*, 2006, **23**, 10, 2317.

3.  D. Bahl and R.H. Bogner, *AAPS PharmSciTech*, 2008, **9**, 1, 146.

4.  J. Vercruysse, D. Córdoba Díaz, E. Peeters, M. Fonteyne, U. Delaet, I. Van Assche, T. De Beer, J.P. Remon and C. Vervaet, *European Journal of Pharmaceutics and Biopharmaceutics*, 2012, **82**, 205.

5.  J. Maclean, C. Medina, D. Daurio, F. Alvarez-Nunez, J. Jona, E. Munson and K. Nagapudi, *Journal of Pharmaceutical Sciences*, 2011, **100**, 8, 3332.

6.  M.K. Gupta, Y-C. Tseng, D. Goldman and R.H. Bogner, *Pharmaceutical Research*, 2002, **19**, 11, 1663.

7.  BioPharma Asia, March 2014. *http://biopharma-asia.com/technical-papers/neusilin-polymer-extrudates-for-the-development-of-amorphous-solid-dispersions/* [Accessed January 2015]

8.  A. Gryczke, G.S. Schminke, M. Maniruzzaman, J. Beck and D. Douroumis, *Colloids and Surfaces B: Biointerfaces*, 2011, **86**, 2, 275.

9.  M. Maniruzzaman, M. Rana, J.S. Boateng and D. Douroumis, *Drug Development and Industrial Pharmacy*, 2013, **39**, 2, 218.

10. M.B. Hickey, M.L. Peterson, L.A. Scoppettuolo, S.L. Morrisette, A. Vetter, H. Guzman, J.F. Remenar, Z. Zhang, M.D. Tawa, S. Haley, M.J. Zaworotko and O. Almarsson, *European Journal of Pharmaceutics and Biopharmaceutics*, 2007, **67**, 1, 112.

11. D. McNamara, S. Childs, J. Giordano, A. Iarriccio, J. Cassidy, M. Shet, R. Mannion, E. O'Donnell and A. Park, *Pharmaceutical Research*, 2006, **23**, 8, 1888.

12. V. Trask, W.D.S. Motherwell and W. Jones, *International Journal of Pharmaceutics*, 2006, **320**, 1–2, 114.

13. B. Aakeroy, N.R. Champness and C. Janiak, *CrystEngComm*, 2010, **12**, 1, 22.

14. A.V. Yadav, A. Shete, A. Dabke, P. Kulkarni and S. Sakhare, *International Journal of Pharmaceutical Sciences*, 2009, **71**, 4, 359.

15. A. Alhalaweh and S.P. Velaga, *Crystal Growth & Design*, 2010, **10**, 8, 3302.

16. S. Aher, R. Dhumal, K. Mahadik, A. Paradkar and P. York, *European Journal of Pharmaceutical Sciences*, 2010, **41**, 5, 597.

17. L. Padrela, M.A. Rodrigues, S.P. Velaga, H.A. Matos and E.G. de Azevedo, *European Journal of Pharmaceutical Sciences*, 2009, **38**, 1, 9.

18. R.S. Dhumal, A.L. Kelly, P. York, P.D. Coates and A. Paradkar, *Pharmaceutical Research*, 2010, **27**, 2725.

19. A.L. Kelly, T. Gough, R.S. Dhumal, S.A. Halsey and A. Paradkar, *International Journal of Pharmaceutics*, 2012, **426**, 1–2, 15.

20. H. Moradiya, M.T. Islam, G.R. Woollam, I.J. Slipper, S. Halsey, M.J. Snowden and D. Douroumis, *Crystal Growth & Design*, 2013, **14**, 189.

21. H.G. Moradiya, M.T. Islam, M. Maniruzzaman, B.Z. Chowdhry, S. Halsey and D. Douroumis, *CrystEngComm*, 2014, DOI:10.1039/c3ce42457j.

22. K. Boksa, A. Otte, R. Pinal. *Journal of Pharmaceutical Sciences*, 2014, **103**, 2904.

23. *Quality Guideline for Pharmaceutical Development, Q2 (R2)*, International Conference on Harmonisation, US Food and Drug Administration, Federal Register, 71 (106), 2006.

24. *Quality Guideline for Pharmaceutical Development, Q2 (R2)*, International Conference on Harmonisation, US Food and Drug Administration, 74 (66), 2009.

25. H. Wu and M.A. Khan, *Journal of Pharmaceutical Sciences*, 2010, **99**, 3, 1516.

26. L. Saerens, L. Dierickx, B. Lenain, C. Vervaet, J.P. Remon and T. De Beer, European *Journal of Pharmaceutics and Biopharmaceutics*, 2011, **77**, 158.

27. C. Kulkarni, A. Kelly, J. Kendrick, T. Gough and A. Paradkar, *Crystal Growth & Design*, 2013, **13**, 5157.

28. E. Horosanskaia, A. Seidel-Morgenstern and H. Lorenz, *Thermochimica Acta*, 2014, **578**, 74.

# 5 Continuous Co-crystallisation of Poorly Soluble Active Pharmaceutical Ingredients to Enhance Dissolution

Ming Lu

## 5.1 Introduction

Co-crystallisation is considered to be an effective method to modify the physicochemical properties of active pharmaceutical ingredients (API) without changing their physiological action, such as solubility, bioavailability, hygroscopicity, mechanical properties [1–6]. Co-crystal can be defined as crystalline materials that comprise two or more components that are solids at room temperature (to distinguish them from hydrates and solvates) held together by non-covalent interactions [7–9]. It can be prepared by a solvent-based method and a solid-state method. The solvent-based method includes slurry conversion, solvent evaporation, cooling crystallisation and precipitation. The extensive use of organic solvent is a big drawback of the solvent method. The solid-state method involves neat grinding, liquid-assisted grinding and melting method. The advantages and disadvantages of the two methods are summarised in **Table 5.1**. The most significant limitation is the scale-up of these conventional methods.

| Table 5.1 Advantages and disadvantages of three methods for co-crystal manufacture | | |
|---|---|---|
| | **Advantages** | **Disadvantages** |
| Solvent method | • Good control of chemical purity and material properties<br><br>• Ease of reproducibility | • A large number of experiments are necessary to measure the ternary phase diagram<br><br>• Possible significant difference in the solubility of the API and the co-former<br><br>• Environmental pollution<br><br>• Risk of producing solvate during solvent removing<br><br>• Difficult to scale-up |

| Table 5.1 **Continued** | | |
|---|---|---|
| | **Advantages** | **Disadvantages** |
| Grinding method | • Need little or no solvent<br><br>• Produces phases that cannot be formed by the solvent method | • Difficult to scale-up |
| TSE method | • Free of solvent<br><br>• Continuous production<br><br>• Ease of scale-up<br><br>• Good mechanical properties | • Chemical purification is not possible |
| TSE: Twin-screw extrusion | | |

In 2009, twin-screw extrusion (TSE), was firstly reported to continuously manufacture co-crystal of AMG 517 and caffeine [10]. It should be mentioned that hot-melt extrusion (HME) was expressed as TSE in this chapter, because several crystallisation processes in the extruder were solid-state reactions and occurred at the processing temperatures under the melting temperature(s) ($T_m$) or eutectic point of reactants. The TSE method has many advantages over traditional solvent and grinding methods. Firstly, it is a scalable, continuous and solvent-free process to manufacture co-crystal. For example, producing 71 g carbamazepine (CBZ)-nicotinamide (NCT) co-crystal by the solvent method approximately required 1 L organic solvent [11], while extruding 140 g AMG 517-sobic acid co-crystal using a 16 mm twin-screw extruder only took about 4.25 h and did not need any solvent [12]. Another example is the successful scale-up of caffeine-oxalic acid co-crystal (100 g scale) using the mixing screw design at a processing temperature of 75 °C, feed-rate of 5% and screw speed of 100 rpm [13]. Secondly, co-crystal produced by TSE exhibits improved physicochemical and mechanical properties compared with co-crystal produced by the solvent method, such as faster dissolution, higher density, better flowability and compressibility. For example, ibuprofen (IBU)-NCT co-crystal was extruded in the form of spherical agglomerates, which were directly compressible, and thus negated the need for further size modification steps [14]. The mechanical properties of AMG 517-sobic acid co-crystal prepared by TSE and solvent methods were well studied [12]. The TSE sample had higher density and better flowability (classified as easy flowing with 4<flow function coefficient<10) compared with co-crystal produced by solvent method (classified as very cohesive with 1<flow function coefficient<2). These improved mechanical properties of co-crystal allows the TSE method to claim unique superiority, especially when used to prepare high strength tablets with increased drug loads using dry granulation. Moreover, TSE samples had a highly corrugated surface and a larger surface area (2.9 m²/g) than co-crystal produced by the solvent method (0.9 m²/g) resulting from the mechanical action in extrusion (**Figure 5.1**).

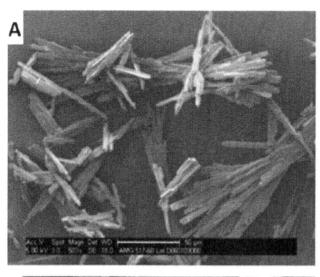

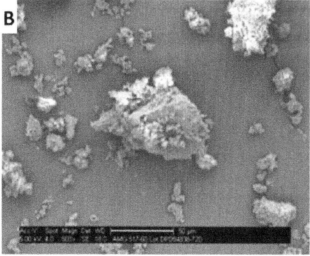

**Figure 5.1** SEM micrographs of AMG 517-sobic acid co-crystal produced by (A) solvent method and (B) TSE method. Reproduced with permission from D. Daurio, K. Nagapudi, L. Li, P. Quan and F. Alvarez-Nunez, *Faraday Discussions*, 2014, DOI: 10.1039/C3FD00153A. ©2014, Royal Society of Chemistry [12]

For CBZ-*trans*-cinnamic-acid co-crystal, the TSE sample demonstrated a faster dissolution than the solvent method sample [15] (**Figure 5.2A**). A similar phenomenon was observed in CBZ-saccharin (SAC) co-crystal (**Figure 5.2B**) [16]. The author suggested that the particle size distribution and purity of CBZ-SAC co-crystal were not the key factors resulting in the rapid dissolution behaviour of the TSE sample. It is well known that large surface area favours fast dissolution. Based on the above-mentioned results of AMG 517-sobic acid co-crystal, the faster dissolution of CBZ-SAC co-crystal produced by TSE might be attributed to the rougher surface and larger surface area, compared with the co-crystal produced by the solvent method.

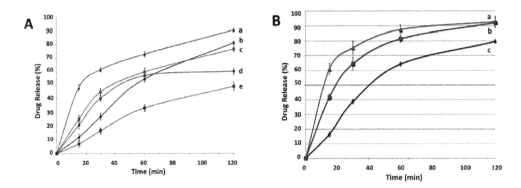

**Figure 5.2** Dissolution profiles of (A) CBZ-*trans*-cinnamic-acid system:
a) co-crystal produced by TSE at 135 °C; b) co-crystal produced by solvent method; c) co-crystal produced by single-screw extrusion at 135 °C; d) co-crystal produced by single-screw extrusion at 125 °C; and e) and pure CBZ. Reproduced with permission from H.G. Moradiya, M.T. Islam, S. Halsey, M. Maniruzzaman, B.Z. Chowdhry, M.J. Snowden and D. Douroumis, *CrystEngComm*, 2014, **16**, 17, 3573. ©2014, RSC [15]. (B) CBZ-SAC system: a) co-crystal produced by TSE at 135 °C and 5 rpm; b) co-crystal produced by TSE at 135 °C and 10 rpm; and c) co-crystal produced by solvent method (pH 1.2, *n* = 3). Reproduced with permission from H. Moradiya, M.T. Islam, G.R. Woollam, I.J. Slipper, S. Halsey, M.J. Snowden and D. Douroumis, *Crystal Growth & Design*, 2014, **14**, 1, 189. ©2014, American Chemical Society [16]

However, TSE-co-crystallisation also has its drawbacks. For example, chemical impurities could be rejected during co-crystallisation in solution, while this chemical purification is impossible with the TSE method. Moreover, the purity of co-crystal might be more difficult to control for the TSE method compared with the solvent method. Fortunately, many researches have focused on how to improve the extent of conversion to co-crystal during TSE.

Generally, TSE is still an efficient, scalable and green method to produce co-crystal for dissolution and bioavailability enhancement. In this chapter, the research progress in continuous co-crystallisation by TSE was reviewed, focusing on the mechanism and critical parameters influencing the extent of conversion to co-crystal.

## 5.2 Mechanism of Continuous Co-crystallisation by Twin-screw Extrusion

The mechanism of continuous co-crystallisation by TSE was poorly documented, while the mechanism of grinding method was well investigated [17–19]. Both of the two methods belong to mechanochemical synthesis. The most remarkable differences between them are the controllable temperature, weaker shear intensity and shorter residence time of TSE compared with the grinding method. As there are many similarities between the two methods, the research findings of the grinding method might give more insight into the mechanism of TSE-co-crystallisation.

From a thermodynamic viewpoint, co-crystallisation process can be considered as the breakage of intermolecular interactions between similar molecules and the formation of new interactions between API and co-former molecules. Maheshwari and co-workers [20] calculated the free energy changes of CBZ-NCT and CBZ-SAC co-crystallisation processes using solubility and solubility product ($K_{sp}$). The negative free energy changes indicated that co-crystallisation revealed the spontaneity of the co-crystallisation processes. The CBZ-NCT and CBZ-SAC equimolar mixtures spontaneously converted to the corresponding co-crystals after storage at 45 °C/0% relative humidity (RH) or 45 °C/75% relative RH for 3 months without any mechanical stress. However, this spontaneous process was extremely slow because the intermolecular contacts were highly limited in the solid-state, which is the prerequisite for intermolecular interaction and mass transfer.

Just as suggested by Rothenberg and co-workers [21], the formation of a liquid phase is essential to facilitate intermolecular contacts and mass transfer. Actually, the intermolecular contact can be mediated by liquid, disorder/amorphous phase or vapour. Correspondingly, the mechanism of grinding co-crystallisation can be divided into a liquid-mediated mechanism, an amorphous state-mediated mechanism and a vapour-mediated mechanism. For TSE-co-crystallisation, only the liquid-mediated mechanism and the amorphous state-mediated mechanisms have been reported. The intermediate of liquid can be further divided into eutectic melts and solvent (including water and organic solvent).

From a dynamic viewpoint, the high temperature and shear force provided by TSE, would accelerate the spontaneous conversion to co-crystal. From a macroscopic perspective, acceleration of the co-crystallisation process at a high temperature can be predicted according to a time-temperature superposition or Arrhenius law. From a microscopic view, high temperature provides energy for molecular mobility and thermal motion of atoms, which weakens the intermolecular interactions between similar molecules. Furthermore, if the processing temperature is higher than the eutectic point or $T_m$ of the components, the newly formed melt should dramatically accelerate the co-crystallisation process due to the highly improved molecular mobility [22]. Besides controllable temperature, shear force provided by TSE was another important factor to promote the formation of co-crystal. For liquid-mediated co-crystallisation, shear force promotes the conversion to co-crystal through improving molecular diffusion, accelerating nucleation and continuously exposing new surfaces for further reaction [19]. For amorphous phase-mediated co-crystallisation, shear force tends to induce disorder or an amorphous phase on the surface of crystal with high mobility. However, there is a lack of evidence to support an amorphous phase-mediated mechanism in TSE-co-crystallisation.

### 5.2.1 Eutectic-mediated Co-crystallisation

#### 5.2.1.1 Ibuprofen-Nicotinamide Case

Dhumal and co-workers [14] suggested that extruding above eutectic point was required for co-crystal formation due to the generation of intermediate melt phase for mass transfer. IBU-NCT was selected as a model pair because it is known to form a eutectic. The $T_m$ of IBU and NCT were 79 and 128 °C, respectively. On heating the equimolar mixtures of IBU and NCT, a eutectic peak appeared around 74 °C. Melting of newly formed co-crystal also occurred at around 90 °C. Three processing temperatures were selected as 70, 80 and 90 °C. When extruded at 70 °C, only a small part of co-crystal formed. Both increasing shear intensity and prolonging residence time cannot increase the extent of conversion to co-crystal. As the processing temperature increased to 80 °C (beyond the eutectic point of 74 °C), the formation of co-crystal dramatically increased. At 90 °C, almost all IBU and NCT converted to co-crystal with a high-mixing screw and a slow screw speed of 20 rpm. In this case, extruding above eutectic point really seemed to be an essential condition for co-crystallisation. Obviously, manufacturing IBU-NCT co-crystal by TSE is a eutectic-mediated process.

## 5.2.1.2 Carbamazepine-Saccharin Case

CBZ-SAC co-crystal is a typical system for mechanochemical co-crystallisation. As mentioned above, the free energy changes of the co-crystallisation process at 25 °C were calculated to be -5.1 kJ/mol [20], confirming the spontaneity of the conversion process under ambient conditions. Mechanical activation, high temperature, humidity and high level of molecular contact might kinetically accelerate this spontaneous conversion process. Jayasankar and co-workers [23] investigated the co-crystal formation of CBZ-SAC by grinding at different temperatures. After cryogenic grinding for 30 min, large amounts of amorphous material with little newly formed co-crystal and unreacted materials were obtained, indicating the amorphous phase was stabilised by lowering the molecular mobility at low temperature and consequently depressing the co-crystal formation. After grinding for 30 min under ambient conditions, approximately 85% conversion to co-crystal was achieved due to higher molecular mobility at the higher temperature compared with cryogenic grinding.

Halasz and co-workers [18] also investigated the mechanism of forming CBZ-SAC co-crystal by grinding using *in situ* synchrotron wide-angle X-ray diffraction (WAXD). The disappearance of the reactant's diffraction peaks was observed with no signs of co-crystal appearing during grinding under ambient conditions, indicating the amorphisation of reactants and no formation of co-crystal (**Figure 5.3A and C**). The different results fort grinding-induced CBZ-SAC co-crystallisation under ambient condition in the two laboratories might be attributed to the mechanical difference between the two milling devices. However, both of the results indicated that producing CBZ-SAC co-crystal by grinding might be an amorphous phase-mediated process.

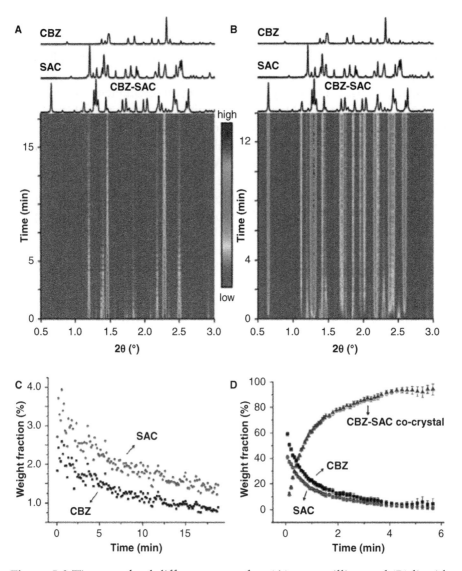

**Figure 5.3** Time resolved diffractograms for: (A) neat milling and (B) liquid-assisted grinding of CBZ and SAC. (C) Time dependent change in phase scale factors for the neat milling of CBZ and SAC. (D) Change in the weight fraction of reactants and products in the liquid-assisted grinding synthesis of CBZ and SAC. Reproduced with permission from I. Halasz, A. Puškarić, S.A.J. Kimber, P.J. Beldon, A.M. Belenguer, F. Adams, V. Honkimäki, R.E. Dinnebier, B. Patel and W. Jones, *Angewandte Chemie*, 2013, **125**, 44, 11752. ©2013, John Wiley & Sons [18]

Daurio and co-workers [13] prepared CBZ-SAC co-crystal by using TSE and grinding methods. There was no co-crystal formed by TSE at a processing temperature of 50 °C and a screw speed of 25 rpm, while co-crystal formation was observed after grinding for 30 min with a final temperature of 45 °C. The author ascribed the failure of co-crystallisation by TSE to the weak shear intensity and short residence time. When the processing temperature of TSE was increased to 190 °C (beyond the $T_m$ of co-crystal, 178 °C), the extent of conversion to co-crystal reached approximately 95%, calculated using [13]C-solid-state nuclear magnetic resonance (SSNMR) spectroscopy. CBZ-SAC was selected as a model system in this work because its co-crystal formation in a grinding process is based on an amorphous phase-mediated mechanism. However, the formation of CBZ-SAC co-crystal by TSE and grinding method seems to have different co-crystallisation mechanisms.

Moradiya and co-workers [16] real-time monitored the CBZ-SAC co-crystallisation process in the extruder by placing a high temperature near-infrared (NIR) probe in different zones of the extruder barrel. When processing at 135 °C (higher than the first eutectic point of CBZ and SAC, 128 °C), almost all CBZ and SAC directly converted to co-crystal form without the detection of an amorphous state (**Figure 5.4**).

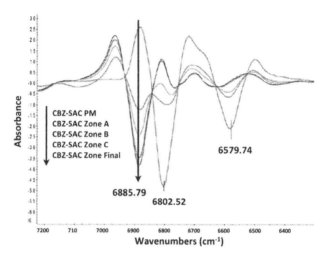

**Figure 5.4** Second derivative of in line NIR spectra of CBZ-SAC system in the mixing zones (A, B, C), the TSE (5 rpm) extrudates, and the physical mixture (PM). Each scan is the average of three different batches. Reproduced with permission from H.G. Moradiya, M.T. Islam, G.R. Woollam, I.J. Slipper, S. Halsey, M.J. Snowden and D. Douroumis, *Crystal Growth & Design*, 2014, **14**, 1, 189. ©2014, American Chemical Society [16]

Dong and co-workers [24] traced the formation of CBZ-SAC co-crystal during the heating process using *in situ* synchrotron WAXD (**Figure 5.5**). The diffraction of co-crystal continuously increased, accompanied by a continuous decrease in the signals of the reactants. No amorphous state was detected during the whole heating process. In addition, the diffraction rings of CBZ-SAC co-crystal appeared with the first eutectic formation, enhanced with the second eutectic formation and maximised during the melt process of newly formed co-crystal (**Figure 5.6**). Obviously, the formation of CBZ-SAC co-crystal during the heating process is based on a eutectic-mediated mechanism.

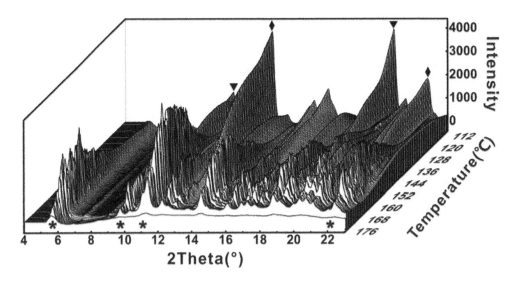

**Figure 5.5** One-dimensional WAXS patterns of CBZ-SAC physical mixture at heating rate of 2 °C/min. ✱: CBZ-SAC co-crystal; ▲: CBZ; and ◆: SAC

Generally, the real-time experiments confirmed two different mechanisms for the formation of CBZ-SAC co-crystal. For the melting method, the spontaneous conversion process was accelerated by the high temperature based on a eutectic-mediated mechanism. For the grinding method, the conversion process was facilitated by mechanical force based on an amorphous phase-mediated mechanism. TSE can be considered as a combination of high temperature and shear force. The failure of co-crystal formation when extruded at 50 °C [13] indicated that the shear force in TSE was too weak to produce amorphous reactants and thus promote the formation of CBZ-SAC co-crystal. Therefore, eutectic formation might be the dominant mechanism in manufacturing CBZ-SAC co-crystal by TSE.

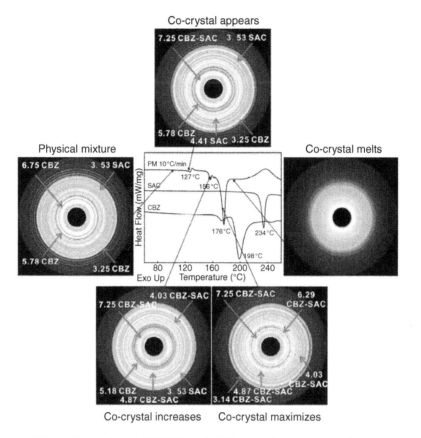

**Figure 5.6** Two-dimensional WAXS and differential scanning calorimetry (DSC) results of CBZ-SAC PM during heating process at 10 °C/min

### 5.2.2 Solvent-assisted Co-crystallisation

Solvent-assisted grinding has been well documented [19, 25–29]. In the above-mentioned CBZ-SAC case in **Section 5.2.1**, the addition of 50 μL of acetonitrile surprisingly accelerated the co-crystallisation of CBZ and SAC during grinding, yielding almost 0.5 g co-crystal within 4 min (**Figure 5.3B and D**).

Substantial improvement in kinetics of co-crystal formation by TSE was also observed by adding catalytic amounts of solvent. For CBZ-SAC case, complete conversion to co-crystal can be achieved by neat TSE at a processing temperature of 135 [16] and 190 °C [13]. After adding about 5 ml of water to 10.4 g of CBZ-SAC equimolar mixture, co-crystal can be extruded at 100 °C with a conversion extent of 87%. On adding ethanol and poly(4-methyl 1-pentene), the co-crystal was extruded at 20~80 °C [13]. The acceleration of co-crystal formation at low temperature should be attributed to the increasing conformational freedom and enhanced opportunities for molecular contact and nucleation.

Solvent-assisted TSE also accelerates the formation of theophylline-citric acid co-crystal [13]. The replacement of citric acid anhydrous (CA) with citric acid monohydrate (CM) dramatically enhanced the extent of the co-crystallisation at various processing temperatures, confirmed by $^{13}$C-cross-polarisation/magic-angle spinning (CP/MAS) spectra (**Table 5.2** and **Figure 5.7**). Furthermore, the lower processing temperature permitted a higher level of conversion to co-crystal. As the processing temperature decreased from 153 to 50 °C, the extent of conversion to co-crystal enhanced to almost 100%. It might be because the water released from hydrates acted as a medium for molecular contact and thus promoted the co-crystal formation. At the lower temperature, the released water could stay for a longer time in the extruder to intensify the water-mediated co-crystallisation. More surprisingly, adding catalytic amounts of water or ethanol could facilitate the formation of pure anhydrous co-crystal (AC) even at a very low temperature of 20 °C.

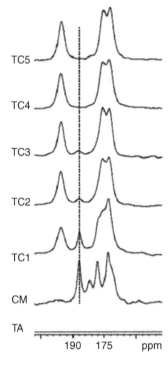

**Figure 5.7** $^{13}$C-CP/MAS spectra of theophylline-citric acid system. TA: Theophylline anhydrous and TC: theophylline-citric acid. Reproduced with permission from D. Daurio, C. Medina, R. Saw, K. Nagapudi and F. Alvarez-Nunez, *Pharmaceutics*, 2011, **3**, 3, 582. ©2014, Multidisciplinary Digital Publishing Institute [13]

The two cases indicate that a catalytic amount of liquid can dramatically improve the efficiency of the co-crystallisation by TSE and depress the processing temperature required to form the co-crystal in the extruder through the mechanism of providing a medium to facilitate molecular diffusion or by templating the formation of a multicomponent inclusion framework [19]. Solvent-assisted TSE is a really promising method for manufacturing co-crystal, especially for heat-sensitive API and co-former.

| | Table 5.2 Outcome of TSE and grinding experiments performed with TC system | | | | | | |
|---|---|---|---|---|---|---|---|
| Lot# | API | Co-former | Type of extrusion | Tp (°C) | Product | Type of grinding* | Product |
| TC1 | TA | CA | Neat | 153 | AC + unreacted | Neat | AC |
| TC2 | TA | CM | Neat | 153 | AC + unreacted | Neat | HC |
| TC3 | Theophylline monohydrate | CA | Neat | 153 | AC + unreacted | Neat | HC |
| TC4 | TA | CM | Neat | 50 | AC | Neat | HC |
| TC5 | TA | CA | Ethanol-assisted | 20 | AC | — | — |
| TC6 | TA | CM | Neat | 20 | HC | — | — |
| TC7 | TA | CM | Water-assisted | 20 | HC | — | — |

*All of the grinding processes were carried out under ambient conditions

HC: Hydrated co-crystal

XRPD: X-ray powder diffraction

The unreacted starting material was observed by XRPD

Reproduced with permission from D. Daurio, C. Medina, R. Saw, K. Nagapudi and F. Alvarez-Núñez, *Pharmaceutics*, 2011, **3**, 3, 582. ©2011, Multidisciplinary Digital Publishing Institute [13]

## 5.2.3 Amorphous Phase-mediated Co-crystallisation

### 5.2.3.1 AMG 517-Sorbic Acid Case

The occurrence of AMG 517-sorbic acid co-crystallisation was temperature independent, but the extent of conversion increased with the increasing temperature (**Figure 5.8**) [12]. At a very low temperature of 10 °C, about 57% conversion to co-crystal occurred using a mixing design screw, which was quantitatively confirmed by $^{19}$F-SSNMR. However, complete conversion only occurred at temperatures higher than 115 °C. Only two eutectic points appeared in the phase diagram: 131 °C between sorbic acid-co-crystal and 194 °C between AMG 517-co-crystal. No eutectic point between AMG 517 and sorbic acid was observed. As the two eutectic points were higher than the processing temperatures used in that study, the author suggested that the formation of co-crystal in the extruder was not mediated by eutectic formation. As the vapour pressure of these compounds does not lend them to sublimation, the vapour-mediated mechanism was also ruled out. Therefore, amorphous phase-mediated co-crystallisation was suggested to be the most likely mechanism driving co-crystal formation in AMG 517-sorbic acid case by TSE. But no direct evidence was given.

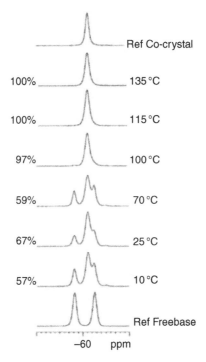

**Figure 5.8** $^{19}$F-SSNMR data for extruded AMG 517-sorbic acid samples as a function of temperature. The numbers shown in percentages refer to the percentage conversion to the co-crystal. Reproduced with permission from D. Daurio, K. Nagapudi, L. Li, P. Quan and F. Alvarez-Nunez, *Faraday Discussions*, 2014, DOI:10.1039/C3FD00153A. ©2014, Royal Society of Chemistry [12]

## 5.3 Critical Parameters Influencing Continuous Co-crystallisation by Twin-screw Extrusion

### 5.3.1 Processing Temperature

Processing temperature is a very important factor influencing continuous co-crystallisation by TSE. According to the dependence on processing temperature, TSE-co-crystallisation can be divided into temperature-dependent and temperature-independent cases.

### 5.3.2 Temperature-dependent Case

This kind of case is very sensitive to processing temperatures. The formation of co-crystal only occurs or accelerates when the processing temperature exceeds a specific temperature, such as in the case of NCT-*trans*-cinnamic-acid [13], CBZ-SAC [13] and IBU-NCT [14].

When producing NCT-*trans*-cinnamic-acid co-crystal by TSE [13], the extent of conversion slightly increased as the processing temperature increased from 80 to 100 °C. However, a dramatic acceleration of co-crystal formation appeared when extruded at 110 °C. This crucial temperature exceeded both the eutectic point of NCT-*trans*-cinnamic-acid and the $T_m$ of co-crystal (78.7 and 103.4 °C, respectively, measured by DSC at a heating rate of 10 °C/min in our laboratory). For the IBU-NCT case as mentioned in **Section 5.2.1**, a dramatic improvement of co-crystal formation was observed when the processing temperature increased from 70 to 80 °C (beyond their eutectic point of 74 °C) [14]. These temperature-dependent cases might be based on a eutectic/melt-mediated mechanism.

### 5.3.3 Temperature-independent Case

For temperature-independent cases, co-crystal can be extruded at much lower temperatures than the eutectic point or $T_m$ of the components. For example, caffeine-oxalic acid co-crystal was extruded at 25 and 75 °C without significant difference in co-crystal purity [13]. Theophylline-citric acid co-crystal was extruded at 20 °C [13] with the aid of ethanol/water or when the reactants were hydrates. As mentioned in **Section 5.2**, AMG 517-sorbic acid was extruded at 10 °C with the conversion extent of 58% [12]. For these temperature-independent cases, the mechanism of co-crystal formation was obviously different from eutectic/melt-mediated mechanism and still poorly understood.

### 5.3.4 Screw Design

Screw design is another important factor influencing the extent of conversion to co-crystal by TSE because it can give different shear intensity. Generally, strong shear always favours the formation of co-crystal. The acceleration of shear force for co-crystal formation is probably based on three mechanisms: 1) inducing disorder or amorphous surface for high mobility; 2) accelerating nucleation of co-crystal through producing local density fluctuation; and 3) continuously exposing new surfaces for reaction.

Dhumal and co-workers [14] employed three kinds of screw configurations to evaluate the influence of shear intensity on co-crystal formation of IBU and NCT. Configuration A was completely composed of purely forward conveying elements, representing the minimum shear intensity. Configuration B provided medium shear intensity. Configuration C supplied the highest levels of distributive and dispersive mixing. A high-intensity screw always resulted in a high conversion rate to co-crystal regardless of screw speeds and processing temperatures, which was attributed to the increasing exposure of fresh surfaces and accelerating nucleation by agitation.

The conversion of AMG 517-sorbic acid co-crystal was higher than 99% calculated by NIR data when extruded at 115 °C using a high-mixing screw design. However, the control experiment, only employing conveying elements, did not obtain co-crystal [10]. The caffeine-oxalic acid system also produced the same phenomenon [13]. These results confirmed the critical role of screw design in co-crystal formation by TSE.

### 5.3.5 Screw Speed

Screw speed has a weaker influence on the formation of co-crystal by TSE compared with processing temperature and screw design. Generally, a fast screw speed produces two contradictory results: strong shear force (facilitating the formation of co-crystal) and short residence time (limiting intermolecular contact and the co-crystal formation). In most cases, residence time seems the dominant factor. For example, decreasing screw speed favoured the formation of IBU-NCT co-crystal, but not as strongly as increasing processing temperature or increasing mixing elements in the screw design [14]. Contradictorily, the high screw speed prevented sufficient contact between IBU and NCT in the screw barrel and thus led to a decreasing extent of conversion to co-crystal. Similar phenomena were observed in the caffeine-oxalic acid case [13] and AMG 517-sobic acid case [13]. In some cases, strong shear force at fast screw speed becomes the dominant factor and thus, favours the formation of co-crystal. For example, the extruded CBZ-SAC co-crystal at 135 °C with screw speed of 5 and 10 rpm exhibited normalised heat enthalpies of -107.89 and

-119.80 J/g, respectively [16], measured by differential scanning calorimetry. The higher heat enthalpy of the high-screw speed sample indicated more complete conversion compared with the low-screw speed sample, which was consistent with the XRPD data. Therefore, the influence of screw speed on co-crystal formation is a result of competition between residence time and shear force. The screw speed should be optimised based on the specific case and extrusion conditions.

### 5.3.6 Feed-rate

The effect of feed-rate on the formation of co-crystal in TSE rarely received attention. Only the AMG 517-sobic acid system was employed to investigate this effect using a high-mixing screw (100 rpm, 100 ºC) [12]. The feed-rate was set as 5, 10 and 15%, respectively. The results indicated that the highest extent of conversion to co-crystal was obtained by using the lowest feed-rate of 5%.

## 5.4 Case Study

### 5.4.1 Matrix-assisted Co-crystallisation of Carbamazepine-Nicotinamide System by Twin-screw Extrusion

Not every co-crystal can effectively improve the solubility and dissolution rate of the API. CBZ-NCT co-crystal is a typical example. *In situ* ultraviolet imaging and Raman spectroscopy revealed the spontaneous conversion of CBZ-NCT co-crystal to CBZ dehydrate during dissolution [30]. As the CBZ dehydrate has lower solubility than CBZ-NCT co-crystal and plain anhydrous CBZ (**Figure 5.9**), the theoretical solubility/dissolution enhancement of co-crystal cannot be obtained. Boksa and co-workers [31] reported matrix-assisted co-crystallisation by TSE as a good method to improve the solubility of CBZ-NCT co-crystal. The CBZ-NCT equimolar mixture was blended with Soluplus® at the weight ratio of 80:20 and then extruded at 115 ºC with a long residence time of 20 min. The complete conversion to co-crystal by TSE was confirmed by the disappearance of the reactants' signal in Fourier-Transform infrared spectrum and powder X-ray diffraction patterns. The PM of Soluplus® and CBZ-NCT co-crystal produced by solvent method at weight ratio of 80:20 was used as a reference to clarify the separate effect of the TSE process. A supersaturation effect was observed in the dissolution of both the matrix-assisted TSE sample and the reference PM. However, Soluplus® the TSE sample could maintain a higher supersaturation degree for a longer time than the reference PM. In addition, the dissolution of the TSE sample resulted in a maximal concentration, 5.3 times greater than the reference PM (1.092 mg/mL compared with 0.192 mg/mL). The area under the dissolution curve of the TSE sample is 3.3-fold greater than that of the reference PM (70.3 mg·min/mL compared with 21.4 mg·min/mL). The results showed that TSE process resulted in a significantly increased dissolution rate and degree of solubilisation relative to the reference PM without any changes in formulation. This enhancement was attributed to the higher mixing homogeneity and smaller particle sizes of co-crystal in the TSE sample compared with in the reference PM.

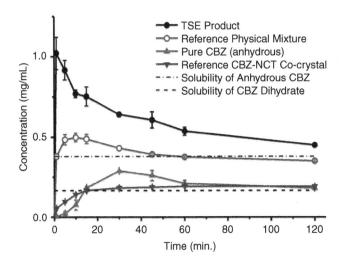

**Figure 5.9** Comparison of the *in vitro* dissolution profiles among CBZ-NCT/Soluplus® formulations. Reproduced with permission from K. Boksa, A. Otte and R. Pinal, *Journal of Pharmaceutical Sciences*, 2014, DOI:10.1002/jps.23983. ©2014, John Wiley & Sons [31]

### 5.4.2 Avoiding Thermal Degradation of Carbamazepine through In Situ Co-crystallising with Nicotinamide

It is known that the extrusion processing temperature should be higher than the $T_m$ of the API for preparing amorphous solid dispersions if there is no strong intermolecular interaction between the API and the polymer carrier. Therefore, thermal degradation of heat-sensitive API is a big problem for HME.

Co-crystallisation with co-former is a good method to modify the $T_m$ of API. Schultheiss [32] reported that 51% co-crystals had moderate $T_m$ between $T_m$ of API and co-former, while 39% co-crystals had lower $T_m$ than $T_m$ of API and co-former. Therefore, Liu and co-workers [33] explored *in situ* forming co-crystal as a single-step, efficient method to depress the $T_m$ of API and thus the processing temperature. The aim was to minimise the thermal degradation of the heat-sensitive drug in extruded amorphous solid dispersions. CBZ and NCT were selected as model API and co-former, respectively, because CBZ-NCT co-crystal has lower $T_m$ (160 °C) than CBZ (190 °C), which is heat-sensitive and decomposes upon melting. CBZ and NIC *in situ* co-crystallised in the polymer carrier during the heating process and then the newly formed co-crystal melted at 160 °C. Consequently, amorphous CBZ-NIC solid dispersions were successfully extruded at 160 °C without any significant thermal degradation of CBZ. The release of CBZ from the CBZ-NCT-polymer solid dispersions was completed within 20 min, faster than from the CBZ-polymer solid dispersion or pure CBZ-NCT co-crystal. *In situ* formation of co-crystal was proven to be an effective approach to prepare chemically stable solid dispersions for heat-sensitive and poorly water-soluble drugs by TSE.

## 5.5 Conclusions

TSE is really a continuous, solvent-free and scalable method to manufacture co-crystal for dissolution enhancement. The extent of conversion to co-crystal strongly depends on the processing temperature and screw design, and moderately depends on the screw speed and feed-rate. Therefore, it is convenient to optimise the extent of conversion to co-crystal through adjusting the processing parameters. Adding a catalytic amount of water or organic solvent is also an efficient method to promote the formation of co-crystal.

However, the research about continuous co-crystallisation by TSE is still at the primary stage. The reported cases are extremely limited and the mechanism is poorly understood. Further investigation is really required.

## References

1.  N. Shan, M.L. Perry, D.R. Weyna and M.J. Zaworotko, *Expert Opinion on Drug Metabolism & Toxicology*, 2014, **10**, 9, 1.

2.  M. Zegarac, E. Leksic, P. Sket, J. Plavec, M.D. Bogdanovic, D.K. Bucar, M. Dumic and E. Mestrovic, *CrystEngComm*, 2014, **16**, 1, 32.

3.  S.L. Childs, P. Kandi and S.R. Lingireddy, *Molecular Pharmaceutics*, 2013, **10**, 8, 3112.

4.  A. Shevchenko, L.M. Bimbo, I. Miroshnyk, J. Haarala, K. Jelínková, K. Syrjänen, B. van Veen, J. Kiesvaara, H.A. Santos and J. Yliruusi, *International Journal of Pharmaceutics*, 2012, **436**, 1, 403.

5.  C. Grossjohann, K.S. Eccles, A.R. Maguire, S.E. Lawrence, L. Tajber, O.I. Corrigan and A.M. Healy, *International Journal of Pharmaceutics*, 2012, **422**, 1–2, 24.

6.  Y.A. Gao, H. Zu and J.J. Zhang, *Journal of Pharmacy and Pharmacology*, 2011, **63**, 4, 483.

7.  G.R. Desiraju, *CrystEngComm*, 2003, **5**, 82, 466.

8.  J.D. Dunitz, *CrystEngComm*, 2003, **5**, 91, 506.

9.  A.D. Bond, *CrystEngComm*, 2007, **9**, 9, 833.

10. C. Medina, D. Daurio, K. Nagapudi and F. Alvarez-Nunez, *Journal of Pharmaceutical Sciences*, 2010, **99**, 4, 1693.

11. A.Y. Sheikh, S.A. Rahim, R.B. Hammond and K.J. Roberts, *CrystEngComm*, 2009, **11**, 3, 501.

12. D. Daurio, K. Nagapudi, L. Li, P. Quan and F. Alvarez-Núñez, *Faraday Discussions*, 2014, DOI:10.1039/C3FD00153A.

13. D. Daurio, C. Medina, R. Saw, K. Nagapudi and F. Alvarez-Núñez, *Pharmaceutics*, 2011, **3**, 3, 582.

14. R.S. Dhumal, A.L. Kelly, P. York, P.D. Coates and A. Paradkar, *Pharmaceutical Research*, 2010, **27**, 12, 2725.

15. H.G. Moradiya, M.T. Islam, S. Halsey, M. Maniruzzaman, B.Z. Chowdhry, M.J. Snowden and D. Douroumis, *CrystEngComm*, 2014, **16**, 17, 3573.

16. H.G. Moradiya, M.T. Islam, G.R. Woollam, I.J. Slipper, S. Halsey, M.J. Snowden and D. Douroumis, *Crystal Growth & Design*, 2014, **14**, 1, 189.

17. D. Gracin, V. Štrukil, T. Friščić, I. Halasz and K. Užarević, *Angewandte Chemie*, 2014, **126**, 24, 6307.

18. I. Halasz, A. Puškarić, S.A.J. Kimber, P.J. Beldon, A.M. Belenguer, F. Adams, V. Honkimäki, R.E. Dinnebier, B. Patel and W. Jones, *Angewandte Chemie*, 2013, **125**, 44, 11752.

19. T. Friscic and W. Jones, *Crystal Growth & Design*, 2009, **9**, 3, 1621.

20. C. Maheshwari, A. Jayasankar, N.A. Khan, G.E. Amidon and N. Rodríguez-Hornedo, *CrystEngComm*, 2009, **11**, 3, 493.

21. G. Rothenberg, A.P. Downie, C.L. Raston and J.L. Scott, *Journal of the American Chemical Society*, 2001, **123**, 36, 8701.

22. K.Chadwick, R. Davey and W. Cross, *CrystEngComm*, 2007, **9**, 9, 732.

23. A. Jayasankar, A. Somwangthanaroj, Z.J. Shao and N. Rodríguez-Hornedo, *Pharmaceutial Research*, 2006, **23**, 10, 2381.

24. P. Dong, L. Lin, Y.C. Li, Z.W. Huang, T.Q. Lang, C.B. Wu and M. Lu, Private Communication.

25. S. Karki, T. Friscic, W. Jones and M.D. Motherwell, *Molecular Pharmaceutics*, 2007, **4**, 3, 347.

26. H.L. Lin, T.K. Wu and S.Y. Lin, *Thermochimica Acta*, 2014, **575**, 313.

27. K.L. Nguyen, T. Friščić, G.M. Day, L.F. Gladden and W. Jones, *Nature Materials*, 2007, **6**, 3, 206.

28. S. Karki, T. Friščić and W. Jones, *CrystEngComm*, 2009, **11**, 3, 470.

29. T. Friščić, S.L. Childs, S.A.A. Rizvi and W. Jones, *CrystEngComm*, 2009, **11**, 3, 418.

30. N. Qiao, K. Wang, W. Schlindwein, A. Davies and M. Li, *European Journal of Pharmaceutics and Biopharmaceutics*, 2013, **83**, 3, 415.

31. K. Boksa, A. Otte and R. Pinal, *Journal of Pharmaceutical Sciences*, 2014, DOI:10.1002/jps.23983.

32. N. Schultheiss and A. Newman, *Crystal Growth & Design*, 2009, **9**, 6, 2950.

33. X. Liu, M. Lu, Z.F. Guo, L. Huang, X. Feng and C.B. Wu, *Pharmaceutical Research*, 2012, **29**, 3, 806.

# 6 Taste Masking of Bitter Active Pharmaceutical Ingredients for the Development of Paediatric Medicines *via* Continuous Hot-Melt Extrusion Processing

Mohammed Maniruzzaman

# 6 Introduction

A significant number of active pharmaceutical ingredient(s) (API) found in oral dosage forms display a bitter taste. The presence of such a taste in drug formulations is undesirable and frequently adversely influences patient compliance. Taste is especially important in pharmaceutical preparations intended for children, who are highly sensitive to taste and actively refuse unpalatable drugs. To date, hot-melt extrusion (HME) has emerged as a novel processing technology for developing molecular dispersions of API using various polymer or/and lipid matrices. This has led to the use of the technique for the production of taste masked, time controlled, modified, extended and targeted drug delivery systems [1, 2]. Pharmaceutical ingredients which can be used to mask the unpleasant and bitter taste of various API are compatible with HME methodologies.

## 6.1 Hot-Melt Extrusion as an Active Taste Masking Technique

HME can be advantageous, compared to the other available conventional techniques, for taste masking as has already been described in the literature [1, 2]. It is a continuous non-ambient process which is easy to scale-up, can be used with moisture sensitive actives, provides enhanced API stability within the carrier matrix and is not time consuming. Since the time exposure of the API in the mainstream extrusion process is quite short, relatively low to medium heat sensitive molecules can also be processed *via* HME either for taste masking purposes or to achieve modified-release characteristics.

### 6.1.1 Taste Masking via Continuous Hot-Melt Extrusion Process

HME has only recently been used as a potentially powerful tool for the production of polymeric extrudates with taste masking properties *via* continuous processing.

Different polymeric systems have already been identified and their usefulness has been demonstrated for taste masking of bitter API *via* HME. Despite the fact that there has only been a limited number of taste masking studies using this technique, various researchers have demonstrated that HME can be used effectively for facilitating taste masking purposes, typically by selecting the appropriate inactive ingredients/excipients, as well as optimising the processing parameters [2].

The taste perception of pharmaceutical formulations has a major influence on patient compliance and therefore the taste masking of bitter API is a major challenge especially for the development of orally administered dosage forms in the pharmaceutical industry [3]. In reality, most or many of the active pharmaceutical substances have either an unpleasant taste (such as bitterness) or saltiness. Some of them may also cause oral irritation such as a metallic and/or spicy taste or astringency. For these reasons, the need for a pleasant taste becomes a key aspect for patient palatability [4]. Taste masking of various bitter active ingredients can be carried out using various techniques depending on the type of API and the type of formulation [5].

Currently, various taste masking approaches are available with the most common involving film coating [6] and addition of sugars, flavours or sweeteners [7] which are often limited due to regulatory requirements. Freeze-drying [3]; microencapsulation [7, 8]; fluidised bed coating [3]; high shear mixing [9]; supercritical fluids [8, 9]; particle engineering/coating [10, 11]; and spray drying [12] have been reported to be used as successful techniques for the purposes of taste masking various bitter active substances. In addition to the aforementioned conventional methods, different advanced and chemical taste masking approaches such as drug complexation by cyclodextrines [13, 14]; ion exchange resins [15]; prodrugs and different salt formations [16] have been reported in the literature. Similarly, HME was introduced as an effective tool to manufacture polymeric or lipidic extrudates with improved taste masking properties [17]. Different polymeric systems have been processed for taste masking purposes by using HME in order to apply an inert layer that creates a physical barrier around the drug.

Gryckze and co-workers (2011) successfully masked the taste of ibuprofen (IBU) which was achieved by intermolecular forces (e.g., hydrogen bonding) in the polyelectrolyte complex between the active substance and the polymer matrix, by processing oppositely charged compounds [4]. Similar results were observed by Maniruzzaman and co-workers (2012) where the authors successfully managed to mask the bitterness of paracetamol (PMOL) by applying the same mechanism of intermolecular hydrogen bonding [3]. In addition, solid dispersions where the drug is molecularly dispersed within the polymer matrix have shown effective masking of the drug's unpleasant taste [18]. Successful taste masking requires the development of HME processing conditions, the drug/polymer ratio and the selection of the appropriate formulation components. HME can also be used for the development of robust formulations with increased patient palatability and compliance [18].

Taste masking evaluation of pharmaceutical dosage forms *in vivo* and *in vitro* are usually carried out by human taste panels and the electronic taste sensing systems, respectively. The latter can be employed to determine (structural) differences between different pharmaceutical formulations [3]. Commercially available electronic-tongues such as an Astree e-tongue have been well studied and evaluated for taste masking purposes. The results showed a very good correlation with the findings of the human taste panels, good reproducibility, low detection limits and high sensitivity [19, 20]. Furthermore, in comparison to techniques which evaluate the taste of final extruded formulations (EF) the Astree e-tongue (Alpha MOS, France) has been systematically used to evaluate the bitterness of pure active substances [21]. Therefore, we report for first time the use of an e-tongue to study the taste masking of two polymers with almost identical chemical structures by using two model drugs in order to investigate their effect in cationic drugs.

### 6.1.2 Polymers as Suitable Carriers for Taste Masking

Gryczke and co-workers were the first [22, 23] to evaluate and demonstrate the use of HME as an effective tool to mask the bitter taste of Verapamil hydrochloric acid (VRP). These researchers extruded basic salts of VRP to produce taste masked granules by embedding them in methyl and ethyl methacrylate anionic based copolymer (Eudragit® L100 and Eudragit® L100–55) matrices. They concluded that HME processing using cationic drugs and polymers containing a high concentration of anions, can result in the formation of strong drug-polymer complexes (i.e., polyelectrolytes) during the melt extrusion process, by the addition of a suitable plasticiser (which facilitates extrusion by reducing the processing temperatures). Gryczke and co-workers also claimed that HME processing enabled the formation of strong intermolecular interactions between the active functional groups of the drug and the polymers. In their study, taste masking was achieved with a lower amount of VRP as only a slightly bitter taste at 70% loading was found, while neutral or no taste was detected in the lower loading range (30–50% wt/wt ratios). The strength taste felt in the mouth with respect to the bitterness of the VRP was altered significantly when the samples were processed at higher temperatures and lower screw rotation speeds. The VRP/polymer extrudates revealed possible API-polymer interactions by first layer atomic surface analysis using X-ray photo-electron spectroscopy (XPS). Possible drug-polymer interactions were modelled using computer simulation which, together with the XPS findings, concluded that the interaction occurred *via* a hydrogen bond between the carboxyl group of the polymer and the amide group of the API [24]. Furthermore, the lower N-coefficient values determined from XPS data allowed estimates of the strength of the foregoing interaction between the tertiary nitrogen atom of VRP and the carboxylic groups of the Eudragit® polymers. The lower the N-coefficient values, the higher the number of protonated nitrogen atoms and thus stronger intermolecular interactions occur in the polymeric extrudates [22, 23].

In a similar case study Gryckze and co-workers used another anionic polymer, Preparat® 4135F (dissolves at pH > 7.0), as a carrier to extrude VRP without a plasticiser and found an optimum taste suppression of the extruded materials compared to that of the pure API itself, without interfering with the dissolution properties [24]. The same principle was applied in the case of anionic IBU processed using polymers containing cationic functional tertiary amino groups such as Eudragit® EPO (EPO) [25]. Gryczke and co-workers processed IBU, which is water-insoluble (BCS class II), with EPO co-polymer at various ratios. An *in vivo* panelist's score confirmed that the active EF exhibited effective masking of IBU granules at drug concentrations up to 25%. A mild bitterness was observed by the volunteers at concentrations of 33% or above. The researchers also proposed that the reduced bitterness of IBU was due to the strong ionic drug–polymer interactions and as a result, the drug was molecularly dispersed in the polymer matrix. These claims were confirmed by differential scanning calorimetry (DSC), X-ray powder diffraction (XRPD) and Fourier-Transform infrared (FT-IR) investigations. The solid-state characterisation of IBU using FT-IR proved that IBU acts as a hydrogen donor with the hydrogen bonding acceptor amino group. The deprotonation of the –COOH facilitates the formation of a carboxylate salt resulting in an infrared band at about 1,400 cm$^{-1}$corresponding to the –C-O bonds. The authors also concluded that the interactions resulted in a taste masking effect by the molecular dispersion of IBU during the extrusion process [25].

In another study Gryczke and co-workers (2011) [4] processed up to 40% IBU with only EPO (50%) and talc (10%) and demonstrated sufficient taste masking during the extrusion process. It was found that increasing the concentrations of IBU enhanced the drug–polymer interactions, since IBU has been found to display plasticising effects similar to conventional plasticisers [6]. The presence of a single glass transition temperature ($T_g$) and the absence of an IBU melting endotherm confirmed the complete miscibility of IBU and EPO and the creation of a glassy solution wherein IBU was molecularly dispersed within the EPO, thus facilitating a higher dissolution rate and achieving better taste masking efficiency. The ground, extruded materials were compressed into tablets and compared with commercially available Nurofen® Meltlets Lemon orally disintegrating tablets; the extruded tablets displayed better taste masking efficiency and a 5-fold increase in dissolution of poorly soluble IBU compared to that of Nurofen®.

Recently, Maniruzzaman and co-workers [27] embedded and processed four different bitter cationic API [propranolol hydrochloric acid (PRP), diphenhydramine hydrochloric acid (DPD), cetirizine hydrochloric acid (CTZ) and VRP] with two different grades of anionic polymers [Eudragit® L100 and Eudragit® L100-55 (Acryl-EZE®)]. Results obtained from a panel of six healthy human volunteers demonstrated that the HME EF improved the taste significantly compared to that of the pure API alone. In addition, an *in vitro* evaluation conducted using an Astree e-tongue equipped with seven sensors demonstrated significant taste improvement of the extrudates compared to placebo polymers and the pure API alone. Furthermore, solid-state analysis showed the presence of API in an amorphous or molecularly dispersed state while *in vitro* dissolution showed quite fast release for all drug substances. It was concluded that HME can effectively be used to mask the taste of bitter API by enhancing drug-polymer interactions [27].

More recently Maniruzzaman and co-workers [3] conducted a comparative taste masking study of extruded PMOL with EPO and crosslinked polyvinyl pyrrolidone [Kollidon® VA 64 (VA64)] at different drug loadings (ranging from 30 to 60% wt/wt ratios). The taste evaluation of the developed formulations was carried out by using an Astree e-tongue equipped with seven sensor sets, and the generated data were analysed using multidimensional statistics. Analysis of the data showed significant suppression of the bitter taste of PMOL for both polymers (**Figure 6.1**).

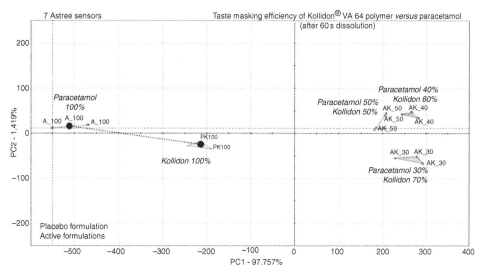

**Figure 6.1** Electronic-tongue 'taste map': PCA of the electrode response between pure PMOL and EF with VA 64 polymer after dissolution for 60 s. PC1: Principal component 1; PC2: principal component 2 and PCA: principal component analysis. Reproduced with permission from M. Maniruzzaman, J.S. Boateng, M.J. Snowden and D. Douroumis, *International Scholarly Research Notices: Pharmaceutics*, 2012, Article ID:436763. ©2012, Hindawi Publishing Corporation [2]

However, PMOL masking was strongly dependent on the nature of the polymeric carriers used and the drug loading in the final formulations. Both polymers showed excellent taste masking for API concentrations up to 50%. Additionally, *in vitro* taste analysis (taste maps) showed significant discrimination between placebo and EF. The three drug-polymer solutions had similar dispersion indices, very different from pure PMOL, indicating a significant taste evolution and an improvement in masking properties for the Kollidon® and EPO extrudates towards pure PMOL. The extruded PMOL formulations were investigated in parallel by *in vivo* taste masking studies which involved a group of six healthy human volunteers. The *in vivo* evaluation was in a good agreement with the e-tongue results. A sensory correlated model based on the partial least square(s) (PLS) method was used to evaluate the correlation with sensory scores. The correlation model was assessed as being valid ($R^2 < 0.8$) despite dispersion and low discrimination between formulations ($p > 0$) [3].

In more recent case studies, Maniruzzaman and co-workers (2012, 2013) [28, 29] extruded a range of four different cationic drugs (DPD, PRP, VRP, CTZ) with three different anionic polymers (Eudragit® L100, Eudragit® S100 and Acryl-EZE®) using 10% triethyl citrate as a plasticiser at 10–20% drug loadings, and characterised the extrudates by analysing the API in the solid-state. The optimised physicochemical processes showed that the drug molecules were dispersed into the polymeric matrices in their amorphous forms. In the aforementioned study, the *in vivo* evaluation showed the taste masking capacity of the polymers in the descending order: Eudragit® L100>Acryl-EZE®>Eudragit® S100. The taste map showed significant discrimination between placebo and active solutions of API. The complex of drug with Eudragit® L100 polymer at 10% drug loading showed a better taste improvement compared to the Acryl-EZE® coating. For the purpose of outlining the possible interaction patterns, depending on the initial orientation of the PRP molecule, the possibility for two hydrogen bonding to be formed were identified after energy optimisation at the B3LYP 6-31G* level using Gaussian 09. After re-optimisation, the H-bonding interaction formed *via* the hydroxyl group was broken due to the presence of chloride ions while the H-bond interaction with the amine group was not affected. Therefore, model calculations indicate that the hydrogen bonds identified in PRP, DPD, Eudragit® L100 and Eudragit® L100-55 formulations were most likely to be formed between the amine groups of the drug molecules and the carbonyl groups of the polymer [28]. Furthermore, significant changes were observed for the chemical shifts in the $^1$H-nuclear magnetic resonance (NMR) spectra of the drug and the drug/polymer formulations. The chemical shift broadening of the $^1$H-NMR signals of EF could be attributed to changes in local magnetic fields arising from intramolecular and intermolecular interactions in the sample. NMR analysis showed significant changes between the T1 relaxation times for the drugs and the drug/polymer extruded matrices. The experimental findings from XPS confirmed one H-bonding pattern as the binding energy peak at ~402.80 eV (higher than typically observed for amines, BE = ~399–400.5 eV and much more for the -NH2+ group) for N 1s is an indication of C-O-NH2+ structure (**Figure 6.2**), whereas the O 1s atom peak at ~534.40 eV confirmed the foregoing conclusion. These results strongly indicate an interaction between the amide group of the API and carboxyl groups of the polymers *via* a hydrogen bond. Dissolution studies of the HME granules showed a rapid release within the first two hours for all API formulated using Eudragit® L100 or Acryl-EZE® [28].

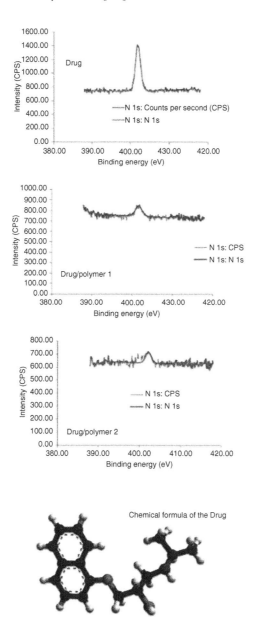

**Figure 6.2** Chemical structure of PRP, N 1s (CPS) binding energy peaks for PRP, PRP/Eudragit® L100 and PRP/Eudragit® L100-55. CPS: Counts per second

### 6.1.3 Use of Lipids as Carriers for Taste Masking in Hot-Melt Extrusion

The use of lipidic matrices in HME to mask the taste of bitter API has recently been successfully employed. Lipidic matrices offer several advantages compared to polymeric matrices for HME processing. Lipid excipients can easily be processed at 20–25 °C above room temperature and sometimes 10–15 °C lower than the melting temperatures of the lipid itself. Therefore extrusion does not need to be plasticiser dependent. Such an approach could be used for processing thermally unstable materials *via* the formation of solid dispersions to achieve taste masking of bitter active substances [30–32]. Breitkreutz and co-workers [32, 33] first introduced 'cold solvent-free extrusion', a technique which utilises different lipids at lower temperatures or even slightly above the room temperature. In their study, sodium benzoate was extruded with either Precirol® ATO5 (glycerol distearate), hard fat or stearic acid in order to mask the taste of the API. Effective taste masking was achieved for at least 5 min in the buccal cavity due to the EPO coating layer which prevented the release of sodium benzoate. In a later study, the same group [34] processed various binary, ternary and quaternary mixtures of powdered lipids with sodium benzoate *via* solvent-free cold extrusion in order to prepare immediate release pellets with solid lipid binders, and compared them to well-known wet extrusion binders, such as microcrystalline cellulose or κ-carrageenan. In this particular study, the authors did not examine the masking efficiency of the manufactured lipid formulations which eventually made the cold extrusion process questionable for taste masking purposes. Therefore, further evaluation of this process is required before optimisation can be achieved for this method to be a viable taste masking approach.

In a similar approach, Michalk and co-workers [35, 36] highlighted the influence of enrofloxacin release on taste perception by assuming that the increase in the amount of drug released raises the probability of taste perception. They proved that the use of a smaller die diameter (0.3 *versus* 0.4 mm) reduces the specific surface area of the milled extrudates and subsequently reduces the amount of API released during short-time intervals (15 s and 1 min). As a result, the investigators concluded that low drug release rates reduce the probability of a bitter taste perception in the mouth. However, no further taste masking efficiency analysis either *in vivo* or *in vitro* was applied in this study.

Recently, Witzled and co-workers [37] developed a continuous, solid-phase lipid extrusion process in order to obtain taste masked granules of praziquantel. In this study, the researchers used various lipids (such as glyceryl tripalmitate, glyceryl dibehenate, glyceryl monostearate, cetyl palmitate and solid paraffin) together with the API, silicon dioxide and polyethylene glycol. The taste masking effect was investigated by using small diameter pellet formulations in a randomised palatability study involving cats. Taste optimisation was based on the fact that animals, such as cats, are sensitive to any bitter tastes and normally tend to reject food that tastes bitter when given, together with bitter tasting medicine. The animals were administrated doses of 5 mg praziquantel in small amounts of dry food and also with canned food. The palatability of the administrated formulations was assessed based on the food uptake by the cats. The results showed 100% food intake for all animals with both dry and canned food which made the lipid-based cold extrusion processing a viable technique for the taste masking of the bitter API.

Furthermore, Breitkreutz and co-workers (2012) applied solid-phase lipid extrusion at room temperature for the taste masked formulation development of the BCS Class II drug NXP 1210 (fenofibrate). In this study, powdered hard fat (Witocan® 42/44 mikrofein), glycerol distearate (Precirol® ATO 5) and glycerol trimyristate (Dynasan® 114) were investigated as lipid binders. The lipid-based formulations were useful for taste masked granules or pellets containing poorly soluble drugs [38, 39]. In the foregoing study, the authors claimed that the dissolution rate from extrudates was significantly increased when compared to pure fenofibrate powder or a physical mixture (PM) of the components. All extrudates were found to be superior to the marketed formulations in terms of taste efficiency and palatability. Furthermore, in the stability study, the researchers did not find any evidence of drug recrystallisation even after 26 weeks of storage for fenofibrate - copovidone extrudates with or without polymeric additives. Both the degree and the duration of supersaturation decreased with increasing storage periods for most of the extrudates. The interesting finding from the study was that by adding polymers with differing release characteristics to the drug–carrier mixture, the dissolution performance of hot-melt extruded taste masked solid dosage forms can be readily adapted to meet specific requirements [38].

## 6.2 Continuous Manufacturing of Oral Films *via* Hot-Melt Extrusion

Films are prepared by three main approaches: solvent casting, spray coating and HME. The solvent casting approach can provide films which are uniform, transparent, and flexible and of different thickness; moreover the films can be produced at low unit cost [39]. There is a limitation in the use of organic based solvents for water-insoluble polymers and difficulty in removing residual solvents, which presents health and environmental concerns [40–43]. These make HME a suitable alternative for producing films. Hot-melt extruded films are prepared by simply blending the relevant film forming polymer, active ingredient and plasticiser prior to feeding through the hopper of the preheated extruder and transferred into the heated barrel by a rotating extruder screw. Generally three main ingredients are required for successful formulation of hot-melt films i.e., film forming polymer(s), active ingredient(s) and plasticiser; [44] the plasticiser is required to maintain film flexibility.

Because films used for oral or trans-mucosal administration inadvertently make contact with the tongue, the taste of bitter API are required to be masked. Several approaches are employed, the most common being addition of sugar based excipients. For example, Mishra and Amin formulated rapidly disintegrating solvent cast films containing CTZ and taste masked using flavours and sweeteners [45]. Although this approach is appropriate, it requires addition of an extra excipient (sugar) that needs optimising. Furthermore, there is always the possibility of an after taste when administered. HME of films therefore provides a unique opportunity to mask the bitter taste of bitter API and is particularly useful for paediatric patients. Different polymeric materials have been employed for this purpose including acrylic and Eudragit® (lidocaine) [46]; hydroxypropyl methylcellulose (ketoconazole) [47]; and hydroxypropyl cellulose (HPC) (hydrocortisone) [48]. Cilurzo and co-workers showed the suitability of maltodextrin (MDX) as a film forming polymer producing fast dissolving hot-melt extruded films of piroxicam (PRX) [49]. The authors concluded in their study that the dissolution behaviour of PRX from the MDX based films is influenced by the wettability of the hydrophilic carrier, the solid-state of the drug (i.e., amorphous or crystalline) and the presence of microcrystalline cellulose [49].

Although the literature is quite extensive for thin films prepared by HME techniques, reports of taste masked films achieved by melt extrusion are scarce. For example, HME muco-adhesive films loaded with clotrimazole intended for local action in the oral cavity have also been reported [50]. The authors concluded that these films showed desirable and optimum bioadhesive and drug release characteristics. In a related study, Prodduturi and co-workers [51] showed that the molecular weight of HPC had significant effects on the mechanical and drug release properties of HME films containing clotrimazole. As discussed previously, the mechanisms involved in taste masking of bitter API in HME solid dosage forms such as tablets (produced by calendering or injection moulding) are expected to play a role in taste masking of bitter API present in the corresponding films since the same polymers are employed in both tablets and films. However, this requires further investigations and testing.

### 6.2.1 Case Study: Taste Masking of Bitter Active Pharmaceutical Ingredients via Continuous Hot-Melt Extrusion Processing

CTZ and VRP are white crystalline powders with a very bitter taste, mainly used as an antihistaminic drug and for treating high blood pressure, chest pain, and irregular heart rhythms, respectively. Both drugs appear to be extremely bitter and conventional masking approaches (e.g., fluidise – bed coating) require increased quantities of excipients. The purpose of this study was to develop and optimise CTZ and VRP based melt extruded granules in order to mask the bitter taste. The extrudates were evaluated by both, *in vivo* and *in vitro* studies where an electronic-tongue analysis

(Alpha MOS, France) was used which captures the global taste profile correlating with the *in vivo* data.

## 6.3 Materials

CTZ and VRP were purchased from Sigma Aldrich (UK). Eudragit® L100 and Eudragit® L100-55 (Acryl-EZE®) were kindly donated by Evonik Pharma Polymers (Germany) and Colorcon Ltd., (UK), respectively. The HPLC solvents were analytical grade and purchased from Fisher Chemicals (UK). All materials were used as received.

### 6.3.1 In Vivo *Taste Masking Evaluation*

*In vivo* taste masking evaluation for pure API, polymers and all active EF was performed in accordance to the Code of Ethics of the World Medical Association (Declaration of Helsinki). Six healthy volunteers of either sex (aged 18–25) were selected (three male and three female) from whom informed consent was first obtained (approved by the Ethics Committee of the University of Greenwich, Ref:UG09/10.5.5.12) and were instructed properly to conduct the test. The equivalent of 100 mg of pure CTZ, VRP or CTZ/VRP extrudates (containing equal amounts of API) was held in the mouth for 60 s and then spat out. The selection of samples was random and in between any two samples, mineral water was used to rinse each volunteer's mouth. The bitterness was recorded immediately according to the bitterness intensity scale from 1 to 5 where 1, 2, 3, 4 and 5 indicate none, threshold, moderate, bitter and strong bitterness.

### 6.3.2 In Vitro *Taste Masking Evaluation: Astree e-tongue*

The assays were realised on Astree e-tongue system equipped with an Alpha MOS sensor set #2 (for pharmaceutical analysis) composed of 7 specific sensors (ZZ, AB, BA, BB, CA, DA, JE) on a 48-position auto sampler using 25 ml beakers. Acquisition times were fixed at 120 s. All the data generated on the Astree system were treated using multidimensional statistics on Alpha Soft V12.3 software. Each solution was tested on Astree e-tongue at least 3 times and 3 replicates were taken into account for the statistical treatment. The sensors were washed frequently after each measurement for 60 s. The sample order used was placebo, API, and the active formulations. The average values between 100 and 120 s were used to build the maps. Astree sensors were cleaned in deionised water between each sample measurement.

### 6.3.3 Sample Preparation for Astree e-tongue

An *in vitro* taste masking evaluation was carried out with an Astree e-tongue equipped with 7 different sensors. To be as close as possible to the panellists' taste conditions, each drug was diluted for 60 s under magnetic stirring in 25 ml of deionised water to reach an API concentration corresponding to a final dose of 100 mg. Then solutions were filtered (60 s) with Buchner funnel fitted with filter paper at 2.5 μm pore size (**Table 6.1**). One single analysis was done for each API.

| Table 6.1 Sample preparation for taste analysis *via* Astree e-tongue | | | | | | |
|---|---|---|---|---|---|---|
| Description | Type | Drug (%) | Polymer (%) | Drug (mg) | Polymer (mg) | Total (mg) |
| VRP | Active | 100 | – | 100 | – | 100 |
| EZE | Polymer | – | 90 | – | 900 | 900 |
| CTZ/EZE | Extruded | 10 | 90 | 100 | 900 | 1,000 |
| CTZ/Eudragit® L100 | Extruded | 10 | 90 | 100 | 900 | 1,000 |
| VRP/EZE | Extruded | 10 | 90 | 100 | 900 | 1,000 |
| VRP/Eudragit® L100 | Extruded | 10 | 90 | 100 | 900 | 1,000 |

## 6.4 Results and Discussion

The theoretical approach derived from the solubility parameter suggests that compounds with similar δ values are likely to be miscible. The reason is that the energy of mixing from intramolecular interactions is balanced with the energy of mixing from intermolecular interactions [29, 52, 53]. Greenhalgh (1999) demonstrated that compounds with $\Delta\delta < 7$ MPa$^{1/2}$ were likely to be miscible and compounds with $\Delta\delta > 10$ MPa$^{1/2}$ were likely to be immiscible [29].

As can be seen in **Table 6.2** the difference between the calculated solubility parameters of the polymers and the drug indicate that both CTZ and VRP are likely to be miscible with both polymers. By using the Van Krevelen–Hoftyzer the $\Delta\delta$ values for CTZ/ Eudragit® L100, CTZ/Acryl-EZE®, VRP/Eudragit® L100 and VRP/Acryl-EZE® are 2.07, 0.97, 3.4 and 2.3 MPa$^{1/2}$, respectively.

| Table 6.2 Solubility parameters calculations summary for both drugs and polymers | | | | |
|---|---|---|---|---|
| Excipients | $\Delta p$ (MPa$^{1/2}$) | $\Delta d$ (MPa$^{1/2}$) | $\Delta h$ (MPa$^{1/2}$) | $\Delta$ (MPa$^{1/2}$) |
| CTZ | 5.42 | 17.50 | 9.60 | 20.68 |
| VRP | 5.12 | 17.31 | 6.95 | 19.35 |
| Eudragit® L100 | 0.41 | 19.31 | 12.03 | 22.75 |
| EZE | 0.25 | 18.22 | 11.69 | 21.65 |

$$R_{a(v)} = \sqrt{\left[\left(\delta_{v2} - \delta_{v1}\right)^2 + \left(\delta_{h2} - \delta_{h1}\right)^2\right]}$$

Furthermore, an advanced two dimensional approach proposed by Bagley and co-workers was also used to predict drug-polymer miscibility as shown in **Table 6.2**. The rationale behind applying the Bagley advanced solubility parameters was to justify the results determined by partial solubility parameter calculations. Studies showed that, in many cases solubility parameters calculated by using the Van Krevelen equation showed a discrepancy in relation with possible drug-polymer miscibility. Maniruzzaman and co-workers [1] have reported a similar study where the authors found that the difference of the calculated partial solubility parameter of PMOL and VA 64 was below 7 MPa$^{1/2}$ but eventually was proved to be immiscible at the end by means of solid-state characterisation (e.g., DSC, XRPD). However, the Bagley advanced parameter provided a more accurate estimation of the possible drug-polymer miscibility, complementing the findings from the solid-state analysis.

The two-dimensional approach can provide a more accurate prediction of the drug-polymer miscibility by calculating the distance Ra(v) using the Pythagorean Theorem [54, 55]. By this approach, two components are considered miscible when Ra(v) ≤ 5.6 MPa$^{1/2}$. **Table 6.2** shows that the hydrogen bonding parameters of the polymers are quite different compared to the dispersion and polar forces represented by the combined solubility parameter δv. Small differences between the corresponding solubility parameters (Δδ) can be observed for all drug/polymer combinations. It can also be concluded that even though the δp values of the drug-polymer combinations differ significantly, it does not affect the predicted miscibility when incorporated within the combined solubility parameter (δv).

Surface morphology was examined by scanning electron microscopy for both the drug and extrudates. The particles size range for all extruded materials varied from 50–200 μm after optimising the milling process. The extrudates containing Eudragit® L100 and Acryl-EZE® exhibited no drug crystals on the extrudate surface with CTZ. Similarly, no VRP crystals were observed on the surface in all drug/polymer extrudates. Furthermore, the sieving analysis (data not shown) presented particle sizes lower than 500 μm for all EF ranging from 40–400 μm.

DSC was used to analyse the solid-state of pure API, polymers, their binary blends and active EF. The overall findings from DSC results are summarised in **Table 6.3**. The DSC thermograms of pure CTZ and VRP showed melting endothermic peaks at 225.6 °C ($\Delta H$ = 104.2 ± 0.5 J/g) and 145.9 °C ($\Delta H$= 106.5 ± 0.5 J/g), respectively. Similarly, the pure polymers showed $T_g$ at 83.97 °C corresponding to Acryl-EZE®, Eudragit® L100-55 and 164.83 °C corresponding to Eudragit® L100, respectively.

| Table 6.3 Summary of DSC results of pure drugs, polymers and formulations | | | |
|---|---|---|---|
| Formulations | $T_g$ (°C) | Melting endotherm/enthalpy (°C /$\Delta H$, Jg$^{-1}$) | |
| CTZ | 55.36 | 225.6/104.2 | |
| VRP | 51.65 | 145.9/106.5 | |
| Eudragit® L100 | 164.38 | – | |
| EZE | 83.97 | 59.2 | |
| PM and EF | | | |
| | PM (°C) | EF (°C /$\Delta H$, Jg$^{-1}$) | PM (°C /$\Delta H$, Jg$^{-1}$) |
| CTZ/Eudragit® L100 | 113.60 | 105.72/6.85 | 222.90/31.23 |
| CTZ/EZE | 73.19 | 64.78/2.64 | 187.55/9.74 |
| VRP/Eudragit® L100 | 113.68 | 74.73/8.15 | 146.41/14.39 |
| VRP/CTZ | 72.98 | 64.11/3.42 | 161.30/11.95 |

A sharp melting peak was observed in the Acryl-EZE® thermogram at 59.2 °C (data not shown), which was attributed to the presence of crystalline plasticisers and other ingredients in the co-processed formulation of Eudragit® L100-55. The same melting peak was also visible in the PM of both drugs. It was observed that all binary physical drugs/polymers blends exhibited endothermic peaks corresponding to the initial substances at slightly shifted temperatures indicating the existence of the drug in its crystalline form. However, the CTZ and VRP endothermic transitions were absent in all EF. The EF exhibited single $T_g$ peaks ranging from 64 to 105 °C indicating the conversion of the crystalline drugs to amorphous during the HME process. Furthermore the presence of a single $T_g$ of the EF between the drug (CTZ at 55.36 °C and VRP at 51.65 °C) and polymers $T_g$ strongly supports the formation of solid molecular dispersions of drugs into the polymer matrices [52–54]. According to the Gordon Taylor equation, the miscible drug/polymers extrudates would exhibit a broad single $T_g$ at the intermediate position of the $T_g$ of amorphous drug and polymers. Nevertheless, the characteristic peak of CTZ and VRP could not be found in the thermograms of the EF, indicating that the EF possess different physicochemical properties from the physical drug/polymer mixtures.

X-ray analysis, studied the drug-polymer extrudates, including pure drugs and PM of the same composition and the diffractograms were recorded to examine both API physical (crystalline or amorphous) states. The diffractograms of pure CTZ and VRP presented distinct peaks ranging from 7.0°, 9.10°, 14.01°, 15.0°, 17.3°, 18.1°, 18.8°, 19.0°, 21.0°, 22.10°, 24.0°, 25.0°, 26.05°, 27.5°, 29.0°, 38.5° 2θ and 2.2°, 4.2°, 5.1°, 9.3°, 12.5°, 18.2°, 19.0°, 19.3°, 23.3°, 25.0°, 28.5°, 31.0°, 35.2°, 36.1°, 37.1° and 39.5° 2θ, respectively. The PM of all formulations presented identical peaks at lower intensities suggesting that both drugs retain their crystalline properties. The PM formulations containing Eudragit® L100-55 for both drugs showed distinct crystalline peaks due to the presence of crystalline plasticiser with co-processed amorphous Eudragit® L100-55 along with the active substances. No intensity peaks were observed in the extruded Acryl-EZE® formulations. Similarly, the absence of intensity peaks for both CTZ/Eudragit® L100 and VRP/Eudragit® L100 was observed in the diffractograms of the EF. The absence of CTZ and VRP intensity peaks indicates the presence of amorphous API in the extruded solid dispersions, which confirms the DSC results.

### 6.4.1 In Vivo/In Vitro *Taste Evaluations*

The masking efficiency of the developed granules was evaluated *in vivo* (approved by University of Greenwich, UK ethics committee) with the assistance of six healthy human volunteers (aged 18–25). The statistical data collected from the *in vivo* study for the pure active substances and the EF are depicted in **Figure 6.3.** The data analysis showed significant suppression ($p < 0.05$) of the bitter taste for both API. These results demonstrate the influence of the polymeric carriers and the importance of the drug loading in the final formulation. Both polymers showed effective taste masking capacities with descending order: Eudragit® L100>Acryl-EZE®. Furthermore, the HME formulations presented an excellent masking effect for active concentrations (10%) of both API. In **Figure 6.3** the sensory data obtained from the panellists showed the taste masking efficiency of Eudragit® L100 was substantially better than Eudragit® L100-55 for both of the API used. This could be attributed to the pH dependant dissolution properties of Eudragit® L100-55 (pH ≥ 5.5) compared to those of Eudragit® L100 (pH ≥ 6) as the saliva of healthy humans has pH ~7.4. However, the sensory scores for both API in all formulations are within the range (below 2), which is considered satisfactory [1].

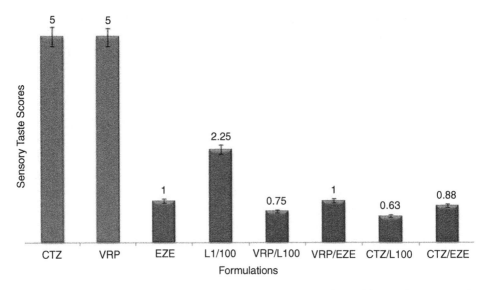

**Figure 6.3** Sensory taste scores of human volunteers for all formulations and pure materials

To evaluate the taste masking *in vitro*, Astree e-tongue data were analysed for both drugs using PCA associated to complementary data processing. The distances between active and polymer samples were calculated as they are indicative of taste masking power of the extrudates. Also, a discrimination index (DI) in % was determined for each solution. This indicator takes into account the average difference between the pairs to compare the solutions of each sample as well. The closer the DI values to 100%, the longer the distance between groups and the lower the dispersion (less masking effects). The DI therefore helps to assess the level of significance of difference between the groups. Sensory correlated models based on PLS were also built to evaluate the correlation with sensory scores. The results are presented in the following sections for each drug.

The taste map in **Figure 6.4** shows significant discrimination between placebo and active samples. Astree sensors are able to detect the presence of the drug in the coated formulations. Focusing on pure VRP and the VPP/Eudragit® L100 (10/90%), EF clearly shows a better taste improvement compared to VRP/Acryl-EZE® samples. The observed distance proximity between the extrudates and pure polymers is about four times greater for VRP/Acryl-EZE® compared to VRP/Eudragit® L100. Similar results can be observed for the pure CTZ and extruded samples.

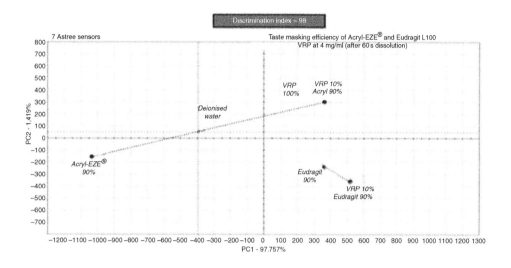

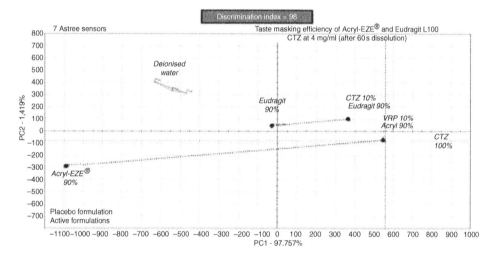

**Figure 6.4** Signal comparison between active and placebo formulations with Eudragit® L100 and Acryl-EZE® extrudates containing CTZ and VRP (dissolution for 60 s)

The masking effect can be also seen in **Figure 6.5** where, CTZ/Eudragit® L100 and VRP/Eudragit® L100 are detected as less bitter compared to pure API, which implies excellent masking improvement (70–72%). On the other hand, the variation observed for the Acryl-EZE® solutions is important especially for the Acryl-EZE® solutions, where a significant difference between the panelist scores and the e-tongue assessment was observed. Eudragit® L100 solutions have a better proximity with a significant improvement of VRP taste due to its higher dissolving pH values (6.0). This trend is likely to be linked with the pH (~5.5) of the distilled water used for the e-tongue studies. Eudragit® L100-55 dissolves in pH > 5.0 and the use of distilled water resulted in the release of increased drug amounts.

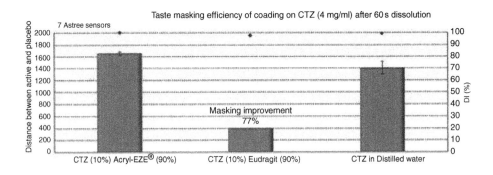

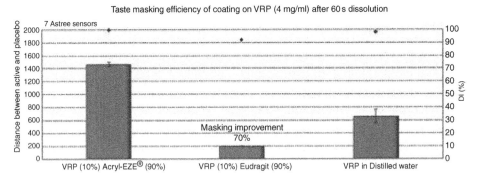

**Figure 6.5** Distance and discrimination comparison between signal of pure CTZ and VRP formulation and each polymer's formulation on Astree e-tongue (after 60 s dissolution)

As shown in PLS, models showed a good correlation. VRP was found to be less bitter compared to CTZ by the panelists and it was confirmed by the e-tongue measurements. Interestingly, Eudragit® L100 was assessed as quite bitter both by the panelists and the e-tongue while the EF showed significant taste suppression. This is an additional indication of possible drug-polymer interactions that lead to taste suppression. The results showed good interpretation of taste masking of all excipients as standard deviation (SD) < 50: Fair, SD < 30: good. The correlation model is considered as valid and fits with panel perception (**Figure 6.5c**; $R^2$ > 0.8). However, it should be considered carefully as the data on sensory tests (number of panelists, variability on measurement) were not communicated. Overall, the taste masking effect of the polymers suggests possible drug-polymer interactions [29] that result in the reduction of the drug bitterness while the active ingredient is molecularly dispersed in the polymer matrix.

## 6.5 Conclusions

HME has proved to be a robust method of producing several drug delivery systems enlarging the scope to include a range of polymers/lipids and API that can be processed with or without plasticisers. HME can also be successfully implemented as an alternative taste masking approach for bitter API. The selection of the appropriate masking agent at the appropriate drug–excipient ratio is a prerequisite for successful taste masking. In addition, processing parameters such as screw speed, zone temperatures and die diameter can lead to optimisation of taste masking methods. In the case study, HME was employed as a robust processing technique for the manufacture of VRP and CTZ taste masked formulations. Both API were found molecularly dispersed in the polymer matrices. The optimised formulations evaluated by *in vivo* and *in vitro* tests to assess taste masking efficacy correlated with each other. The EF of Eudragit® L100 demonstrated better taste masking compared to those of Acryl-EZE® while the e-tongue was found to be a valuable tool for taste masking assessments and formulation development.

## References

1. M. Maniruzzaman, M. Rana, J.S. Boateng and D. Douroumis, *Drug Development and Industrial Pharmacy*, 2013, **39**, 2, 218.

2. M. Maniruzzaman, J.S. Boateng, M.J. Snowden and D. Douroumis, *International Scholarly Research Notices: Pharmaceutics*, 2012, Article ID:436763.

3. M. Maniruzzaman, J.S. Boateng, M. Bonnefille, A. Aranyos, J.C. Mitchell and D. Douroumis, *European Journal of Pharmaceutics and Biopharmaceutics*, 2012, **80**, 2, 433.

4. A. Gryczke, S. Schminke, M. Maniruzzaman, J. Beck and D. Douroumis, 2011, **86**, 275.

5. Z. Ayenew, V. Puri, L. Kumar and A.K. Bansal, *Recent Patents on Drug Delivery & Formulation*, 2009, **3**, 26.

6. D. Douroumis, A. Gryczke and S. Schminke, *AAPS PharmSciTech*, 2011, **12**, 141.

7. M.F. Al-Omran, S.A. Al-Suwayeh, A.M. El-Helw and S.I. Saleh, *Journal of Microencapsulation*, 2002, **19**, 45.

8.  P.P. Shah, R.C. Mashru, Y.M. Rane and A. Thakkar, *AAPS PharmSciTech*, 2008, **9**, 377.

9.  B. Albertini, C. Cavallari and N. Passerini, *European Journal of Pharmaceutical Sciences*, 2004, **21**, 295.

10. J.P. Benoit, H. Rolland, C. Thies and V.V. Vande, inventors; Centre De Microencapsulation, assignee; US6087003, 1994.

11. M.H. Mazen and P. York, inventors; Nektar Therapeutics UK, Ltd., assignee; US7115280, 2006.

12. D. Bora, *AAPS PharmSciTech*, 2008, **9**, 1159.

13. N. Ono, *Journal of Pharmaceutical Sciences*, 2010, **100**, 1935.

14. K. Woertz, C. Tissen, P. Kleinebudde and J. Breitkreutz, *International Journal of Pharmaceutics*, 2010, **400**, 114.

15. R.F. Menegon, L. Blau, N.S. Janzantti, A.C. Pizzolitto, M.A. Corrêa, M. Monteiro and M.C. Chung, *Journal of Pharmaceutical Sciences*, 2011, **100**, 3130.

16. M. Maniruzzaman, J.S. Boateng, B.Z. Chowdhry, MJ. Snowden and D. Douroumis, *Drug Development and Industrial Pharmacy*, 2014, **40**, 145.

17. K. Woertz, C. Tissen, P. Kleinebudde and J. Breitkreutz, *International Journal of Pharmaceutics*, 2011, **417**, 256.

18. K. Woertz, K. Tissen, P. Kleinebudde and J. Breitkreutz, *Journal of Pharmaceutical and Biomedical Analysis*, 2010, **51**, 497.

19. K. Woertz, K. Tissen, P. Kleinebudde and J. Breitkreutz, *Journal of Pharmaceutical and Biomedical Analysis*, 2011, **55**, 272.

20. C.M. Hansen, *Industrial Engineering and Chemical Research Development*, 1969, **8**, 2.

21. P.J. Hoftyzer and D.W. van Krevelen in *Properties of Polymers*, Elsevier, Amsterdam, The Netherlands, 1976.

22. H.U. Petereit, C. Meier and A. Gryczke, inventors; Röhm GmbH & Co. KG, assignee; WO03/072083A2, 2003.

23. D. Douroumis, M. Bonnefille and A. Aranyos in *Pharmaceutical Applications of Hot-Melt Extrusion*, Ed., D. Douroumis, John Wiley and Sons, New York, NY, USA, 2012.

24. A. Gryckze in *Innovative Formulations by Melt Extrusion with EUDRAGIT® Polymers*, Röhm GmbH & Co. KG, Darmstadt, Germany, 2006.

25. H.U. Peterei, C. Meier and A. Gryczke, inventors; Röhm GmbH & Co. KG, assignee; US20060051412A1, 2006.

26. A. De Brabander, G. van den Mooter, C. Vervaet and J.P. Remon, *Journal of Pharmaceutical Sciences*, 2002, **91**, 1678.

27. M. Maniruzzama, J. Boateng and D. Douroumis, *AAPS Journal*, 2011, **13**, S2, T3273.

28. M. Maniruzzaman, J. Pang, A. Mendham, D. Morgan and D. Douroumis, *AAPS Journal*, 2012, **14**, W5324.

29. M. Maniruzzaman, D. Morgan, A. Mendham, J. Pang, M.J. Snowden and D. Douroumis, *International Journal of Pharmaceutics*, 2013, **443**, 1–2, 199.

30. P. Barthelemy, J.P. Laforêt, N. Farah and J. Joachim, *European Journal of Pharmaceutics and Biopharmaceutics*, 1999, **47**, 87.

31. A. Faham, P. Prinderre, N. Farah, K.D. Eichler, G. Kalantzis and J. Joachim, *Drug Development and Industrial Pharmacy*, 2000, **26**, 167.

32. H. Suzuki, H. Onishi, Y. Takahashi, M. Iwata and Y. Machida, *International Journal of Pharmaceutics*, 2003, **251**, 123.

33. J. Breitkreutz, M. Bornhöft, F. Wöll and P. Kleinebudde, *European Journal of Pharmaceutics and Biopharmaceutics*, 2003, **56**, 247.

34. J. Breitkreutz, F. El-Saleh, C. Kiera, P. Kleinebudde and W. Wiede, *European Journal of Pharmaceutics and Biopharmaceutics*, 2003, **56**, 255.

35. J. Krause, M. Thommes and J. Breitkreutz, *European Journal of Pharmaceutics and Biopharmaceutics*, 2009, **71**, 138.

36. A. Michalk, V.R. Kanikanti, H.J. Hamann and P. Kleinebudde, *Journal of Controlled Release*, 2008, **132**, 35.

37. R. Witzleb, V.R. Kanikanti, H.J. Hamann and P. Kleinebudde, *European Journal of Pharmaceutics and Biopharmaceutics*, 2011, **77**, 170.

38. A. Kalivodaa, M. Fischbacha and P. Kleinebuddeb, *International Journal of Pharmaceutics*, 2012, **429**, 58.

39. J.S. Boateng, H.N.E. Stevens, G.M. Eccleston, A. Auffret, M. Humphrey and K.H. Matthews, *Drug Development and Industrial Pharmacy*, 2009, **35**, 986.

40. C.R. Steuernagel in *Aqueous Polymeric Coatings for Pharmaceutical Dosage Forms*, Ed., J.W. McGinity, Marcel Dekker Inc., New York, NY, USA, 1997, p.79.

41. S. Barnhart in *Modified-release Drug Delivery Technology*, Eds., M.J. Rathbone, J. Hadgraft, M.S. Roberts and M.E. Lane, Informa Healthcare, London, UK, 2008, p.209.

42. *ICH Q3C (R3) Impurities: Residual Solvents*, International Conference on Harmonization, Geneva, Switzerland, 2009. *http://www.emea.europa.eu/pdfs/human/ich/028395en.pdf*. [Accessed on October 2012]

43. J.O. Morales and J.T. McConville, *European Journal of Pharmaceutics and Biopharmaceutics*, 2011, **77**, 187.

44. V.S. Tumuluri, M.S. Kemper, I.R. Lewis, S. Prodduturi, S. Majumdar, B.A. Avery and M.A. Repka, *International Journal of Pharmaceutics*, 2008, **357**, 1–2, 77.

45. R. Mishra and A. Amin, *Pharmaceutical Technology*, 2009, **33**, 48.

46. P.K. Mididoddi and M.A. Repka, *European Journal of Pharmaceutics and Biopharmaceutics*, 2007, **66**, 95.

47. S. Prodduturi, R. Manek, W. Kolling, S. Stodghill and M. Repka, *Journal of Pharmaceutical Sciences*, 2005, **94**, 2232.

48. M. Repka, T. Gerding, S. Repka and M.J. McGinity, *Drug Development and Industrial Pharmacy*, 1999, **25**, 625.

49. F. Cilurzo, I.E. Cupone, P. Minghetti, F. Selmin and L. Montanari, *European Journal of Pharmaceutics and Biopharmaceutics*, 2008, **70**, 895.

50. M.A. Repka, S. Prodduturi and S.P. Stodghill, *Drug Development and Industrial Pharmacy*, 2003, **29**, 757.

51.  S. Prodduturi, R.V. Manek and B. Kolling, S.P. Stodgill and M.A. Repka, *Journal of Pharmaceutical Sciences*, 2004, **93**, 3047.

52.  E.B. Bagley, T.P. Nelson and J.M. Scigliano, *Journal of Paint Technology*, 1971, **43**, 35.

53.  X. Zheng, R. Yang, X. Tang and L. Zheng, *Drug Development and Industrial Pharmacy*, 2007, **33**, 791.

54.  D.J. Greenhalgh, A.C. Williams, P. Timmins and P. York, *Journal of Pharmaceutical Sciences*, 1999, **88**, 1182.

55.  J. Breitkreutz, *Pharmaceutical Research*, 1998, **15**, 1370.

# 7 Continuous Manufacturing of Pharmaceutical Products *via* Melt Extrusion: A Case Study

Mohammed Maniruzzaman

## 7 Introduction

Continuous manufacturing is referred to as a process whereby material is simultaneously charged and discharged. In batch manufacturing the product is most frequently tested off-line with a real lack of opportunities for real time monitoring during the operational process. Continuous manufacturing offers numerous benefits such as integrated processing with fewer steps, smaller equipment and facilities, more flexible operation, lower capital costs and rapid development screening over many conditions. Continuous manufacturing also provides scope for extended on-line monitoring, product quality assurance in real-time complying with the standards set by the US Food and Drug Administration.

Hot-melt extrusion (HME) has been developed as a novel technique for the formulation of oral solid dosage forms in pharmaceutical industries in recent years [1]. A wide variety of downstream processing equipment in HME allows the manufacture of various solid dosage forms including pellets, granules, tablets, capsules and films with different pharmaceutical applications. These solid dosage forms can provide sustained-, modified- or targeted-release by controlling both formulation and processing parameters. Despite the fact that initial research developments have focused on the effects of formulation and processing variables on the properties of final dosage forms [2–5], more recent investigations have focused on the use of HME as a novel manufacturing technology for the development of sustained-release (SR) formulations as well as paediatric formulations [6, 7]. HME can also be adopted for use in a continuous manufacturing sphere. Early studies *via* HME processing have described the preparation of matrix mini-tablets which was followed by further investigations into the properties of SR mini-matrices manufactured from ethyl cellulose (EC), hydroxypropyl methycellulose (HPMC) and ibuprofen [8, 9]. Extruded mini-tablets showed a minimised risk of dose dumping and reduced inter- and intra-subject variability. Vegetable calcium stearate was also seen reported in the development of retarded-release pellets used as a thermoplastic excipient processed *via* HME, where pellets with a paracetamol loading of 20% released only 11.54% of the drug after 8 h due to the significant densification of the pellets [10]. A microbicide intravaginal ring was also prepared and developed from polyether urethane elastomers for the sustained delivery of UC781 (a highly potent non-nucleoside reverse transcriptase inhibitor of human immunodeficiency virus [HIV-1]) *via* HME processing [11]. These reported studies so far have yielded many positive results but also many unanswered questions, e.g., the detailed interpretation/description of the intermolecular interactions between the drug and polymers used to develop SR dosage forms and the identity and validity of the manufactured products *via* a continuous process.

The drug release rate is difficult to control in all existing conventional SR tablets, as drug absorption is influenced heavily by its transition rate in the gastrointestinal tract, resulting in wide variations in the oral bioavailability [12–15]. It has been reported that a dosage form can overcome this problem to a certain extent; however, it requires a primitive manufacturing technology [16, 17].

## Case Study

The purpose of this study is to implement HME as a continuous manufacturing technology to develop SR pellets of steroid hormone (SH) that overcomes these biological and technological problems. A formulation development study on the novel SH SR pellets are hereby reported along with the characterisation of the drug distribution onto the pellets and the release mechanism study.

## 7.1 Materials and Method

### 7.1.1 Materials

SH was used as a model drug; EC N10 and EC P7 were used as polymeric carriers in order to conduct this study.

### 7.1.2 Preparation of Formulation Blends and Continuous Hot-Melt Extrusion Processing

SH formulations with EC N10 and EC P7 were mixed properly in a Turbula TF2 mixer (Switzerland) in 100 g batches for 10 min each, prior to the extrusion process. Drug/polymer ratios used were 10–30:90–70 wt/wt for both polymers. Extrusion of all formulations was performed using a EuroLab 16 twin-screw extruder (Thermo Fisher, Germany) equipped with a 2 mm die in 50/130/140/140/140/140/140/140/140/140 °C (from feeding zone → die) temperature profiles. The screw speed maintained for all extrusions was 100 rpm. The extrudates (strands) produced were cut into pellets of 1 mm length using a pelletiser (Thermo Fisher, Germany). Collected pellets were finally ground by using a rotor mill system (Retsch, Germany) with a rotational speed of 400 rpm to obtain granules (<250 μm).

### 7.1.3 Scanning Electron Microscopy/Energy Dispersive X-ray Analysis

The spatial distribution of discreet chemical phases was evaluated using a cold-cathode field-emission gun scanning electron microscope (SEM) (Hitachi SU8030 FEG-SEM, Japan) and Thermo-Noran (USA) energy dispersive X-ray (EDX) system with 30 mm$^2$ UltraDry window and Noran 7 software. The samples were placed on double-sided carbon adhesive tabs and coated with carbon (Edwards 306 high vacuum carbon evaporation) before SEM/EDX analysis. The elemental distribution on the surface of the pellets was investigated using EDX while surface analysis to characterise the morphology of the pellets was evaluated using SEM at a nominal magnification of 1000× (area = 0.127 × 0.095 mm). The accelerating voltage of the incident electron beam was set at 8 kV. This value was selected in order to minimise beam damage to the sample while maintaining adequate excitation. Principal components were extracted from the X-ray maps using Noran 7 (COMPASS) software. X-ray mapping, the conventional method, only shows the elemental distribution while XPhase has the significant advantage that the phase distribution can be obtained by composition of groups of elements from the extracted principal components analysis (PCA). The data acquisition time varies depending on the resolution and noise. The particle distribution on the surface area is characterised on the basis of chemical composition and morphology. XPhase is used to create phase chemical maps from the elemental intensities in the material. In the current study, many of the phases showed a similar set of elements due to the homogeneous distribution of the combined substances and thus it would be difficult to identify the specific phase with standard X-ray maps in the absence of marked elements of the active pharmaceutical ingredient(s) (API). However, when elemental distribution and their relative abundances are combined into phases derived from the PCA, the resulting phase maps are more easily interpreted.

### 7.1.4 Thermal Analysis

The physical state of the pure drug, physical mixtures (PM) and extrudates were examined by using a Mettler-Toledo 823e (Mettler-Toledo, Switzerland) differential scanning calorimeter. Samples were prepared in sealed aluminium pans (2–5 mg) with a pierced lid. The samples were heated at 10 °C/min under nitrogen atmosphere in a temperature range between -25 and 250 °C.

Characterisation of SH in the molten polymeric carrier was assessed using hot stage microscopy (HSM). During testing, an Olympus BX60 microscope (Olympus Corp., USA) with Insight QE camera (Diagnostic Instruments, Inc., USA) was used to visually observe samples, while a FP82HT hot stage controlled by a FP 90 central processor (Mettler Toledo, USA) maintained temperatures at 20–250 °C. Images were captured under visible and polarised light using Spot Advance Software (Diagnostic Instruments, Inc., USA).

## 7.1.5 Powder X-ray Diffraction

X-ray powder diffraction (XRPD) was also used to assess the solid-state (crystalline or amorphous) of the extrudates where samples of pure and loaded API were evaluated using a D8 Advance (Bruker, Germany) in theta-theta mode, Cu anode at 40 kV and 40 mA, parallel beam Goebel mirror, 0.2 mm exit slit, LYNXEYE™ Position Sensitive Detector with 3° opening and LynxIris at 6.5 mm, sample rotation at 15 rpm. The sample was scanned from 5 to 40° 2-theta with a step size of 0.02° 2-theta and a counting time of 0.2 s per step; 176 channels active on the particle size distribution making a total counting time of 35.2 s per step.

## 7.1.6 In Vitro *Drug Release Studies*

An *in vitro* dissolution study was carried out by using a Varian 705 DS dissolution paddle apparatus (Varian Inc., USA) at 100 rpm and 37 ± 0.5 °C. The dissolution medium pH was maintained as 1.2 by using 750 ml of 0.1 M hydrochloric acid for 2 h. After 2 h operation, 150 ml of 0.20 M solution of dehydrogenate sodium orthophosphate was added into the vessel to give the final pH of 6.8 and the temperature equilibrated to 37 °C. At predetermined time intervals, samples were withdrawn for high-performance liquid chromatography assay and replaced with fresh dissolution medium. All dissolution studies were performed in triplicate.

## 7.1.7 Analysis of Drug Release Mechanism

Zero order kinetics, first order kinetics, Hixson–Crowell, Higuchi and Korsmeyer–Peppas models were used for the analysis of the dissolution mechanism taking the rate constant obtained from these models as an apparent rate constant. The drug release patterns from both coated and uncoated tablets were analysed by release kinetics theories [18–22], as follows:

*Zero order kinetics*:

$$F_t = K_o t \tag{7.1}$$

Where $F_t$ represents the fraction of drug released in time $t$ and $K_0$ the apparent release rate constant or zero order release constant.

*First order kinetics*:

$$\ln(1 - F) = -K_1 t \tag{7.2}$$

Where $F$ represents the fraction of drug released in time $t$ and $K_1$ is the first order release constant.

*Higuchi model*:

$$F = K_2 t^{1/2} \tag{7.3}$$

Where $F$ represents the fraction of drug released in time $t$ and $K_2$ is the Higuchi dissolution constant.

*Hixson–Crowell model*:

$$W_0^{1/3} - W_t^{1/3} = K_s t \tag{7.4}$$

Where, $W_0$ is the initial amount of drug in the pharmaceutical dosage form, $W_t$ is the remaining amount of drug in the pharmaceutical dosage form at time $t$ and $K_s$ is a constant incorporating the surface volume relation.

Dividing **Equation 7.4** by $W_0^{1/3}$ and simplifying:

$$(1 - F)^{1/3} = 1 - K_3 t \tag{7.5}$$

Where $F=1- (W_t/W_0)$ and $F$ represents the drug dissolved fraction at time $t$ and $K_3$ is the release constant. When this model is used, it is assumed that the release rate is limited by the drug particle dissolution rate and not by the diffusion that might occur through the polymeric matrix.

*Korsmeyer–Peppas model*:

$$F = K_4 t^n \tag{7.6}$$

Where $K_4$ is a constant incorporating the structural and geometric characteristics of the drug dosage form, $n$ is the release exponent (e.g., first order release when $n = 1$), indicative of the drug release mechanism and $F$ represents the drug dissolved fraction at time $t$. This model is generally used to analyse the release of which the mechanism is not well known or when more than one type of release phenomenon is involved.

## 7.2 Results and Discussion

The prediction of drug/polymers miscibility in all extruded solid dispersions has successfully been achieved by the solubility parameters ($\delta$) using Van Krevelen–Hoftyzer equation [23]. The miscibility of these drug/polymers is achieved when the balance between the energy of mixing released by intermolecular interactions within the drug/polymers and intramolecular interactions within the components, are optimised [24]. The theoretical approach derived from the solubility parameter suggests that compounds with similar $\delta$ values are likely to be miscible. The reason is that the energy of mixing from intramolecular interactions is balanced with the energy of mixing from intermolecular interactions. Greenlagh (1999) demonstrated that compounds with $\Delta dt < 7$ MPa$^{1/2}$ were likely to be miscible and compounds with $\Delta dt > 10$ MPa$^{1/2}$ were likely to be immiscible [25].

The calculated solubility parameter of SH, EC N10 and EC P7 are 22.60, 25.61 and 25.27 MPa$^{1/2}$, respectively. The difference between the calculated solubility parameters of the polymers and the drug indicates that SH is likely to form solid dispersions with both polymers EC N10 and EC P7. By using the Van Krevelen–Hoftyzer equation the $\Delta\delta$ values for SH/EC N10 and SH/EC P7 are 3.01 and 2.67 MPa$^{1/2}$, respectively.

In the current study, the computation of binding strength between the drug and dimeric form of the polymer was carried out with a quantum mechanical (QM)-based molecular modelling approach using commercial software package Gaussian 09. In the QM-based calculations, the total energy of the system is calculated with respect to all atomic coordinates, and thus the sum of electronic energy and repulsion energy between the nuclei and electrons. Because all electrons within the system are explicitly considered, the QM approach is capable of characterising with accuracy, non-bonded interactions, such as hydrogen bonds and charge-charge interactions. Nonetheless, the present approach focuses on the intrinsic strength of H-bonds and charge-charge interactions between the drug molecule and a small fragment of the polymer, which should be the predominant contributor to the drug-polymer miscibility. It does not take consideration of the effect of the full polymeric matrix on drug loading nor the process of polymer chain swelling to absorb the drug molecules, both of which can also play a part in determining drug-polymer miscibility. Thus, the binding energy obtained from our QM calculations should be an underestimate compared to the free energy estimated using the F–H theory.

Because the dimers and the drugs interact predominantly through hydrogen bonds, we interpret the calculated binding energy as a reflection of the strength of hydrogen bonds. Overall, the strength of the interactions is dependent on both the type of donor and acceptor and the number of hydrogen bonds formed between the drug and the polymer (**Figure 7.1**). The calculations showed that the interactions formed between the drug molecule and the protonated carboxylate groups in the polymers significantly high (15–25 kcal/mol) indicating a strong interaction possibility. In addition, the binding energy through one hydrogen bond exhibits about 15.1 kcal/mol while the binding energy through two hydrogen bonds shows 25.3 kcal/mol. Higher binding energy represents more stable drug/polymer intermolecular interactions formed after the extrusion process [12, 13].

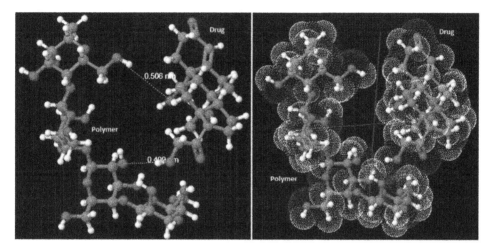

**Figure 7.1** Molecular modelling of drug-polymer interactions. NIR: Near-infrared

### 7.2.1 Continuous Manufacturing of Pellets *via Hot-Melt Extrusion*

Extrusion processing was optimised to obtain the extrudates as white pellets of 1 mm length (**Figure 7.2a**). The temperature profiles, screw speed and the feed-rate were found to be the critical processing parameters for the development of the extruded pellets. During the optimisation of the process to manufacture SR pellets, a range of temperature profiles from 120–140 °C were evaluated and subsequent processing conditions and outputs such as shear force, torque, feed-rate and screw speed were optimised using a design space (design of experiment). Screw configuration was carefully selected in a way to provide intermediate levels of distributive mixing (**Figure 7.2b**). This is typical of the type of screw configurations used in conventional polymer compounding (mixing) operations. Distributive mixing was achieved and optimised in the process during extrusion using a series of bi-lobal mixing/kneading blocks or paddles. These blocks are of a length equal to a quarter of the extruder screw diameter and were arranged at specified angles from the preceding element: 30, 60 or 90° in the forward conveying direction. It is noted that a kneading block arrangement with 30° mixing paddles provides the most forward conveying while 60° provides moderate to less and 90° provides complete mixing with zero forward conveying action.

In an efficient set up of continuous manufacture of any dosage forms, it is paramount to consider and control/monitor quality attributes of the drug for enhanced product performance of interest. In order to monitor critical quality attributes of the drug within the polymeric matrices, an NIR fibre probe as a process analytical technology (PAT) tool was used, and all data generated in-line during the process were collected using appropriate software package provided by Thermo Fisher, Germany. However, the application of NIR as PAT tools for the characterisation will be discussed in a separate manuscript at a subsequent stage. By maintaining the final processing condition and the throughput of the drug/polymer powder blends, the final batch of formulations were extruded continuously at 140 °C with a feed-rate of 1 kg/h and a high screw speed of 100 rpm. Extruded strands were conveyed onto a compressed air facilitated conveyor belt and fed on to the pelletiser attached at the end of the conveyor (**Figure 7.2a**). The pelletiser's speed was adjusted to cut the oncoming strands into pellets of 1 mm length. The length of the pellets was homogenously cut into pieces by the pelletiser as the speed and processing conditions were adjusted.

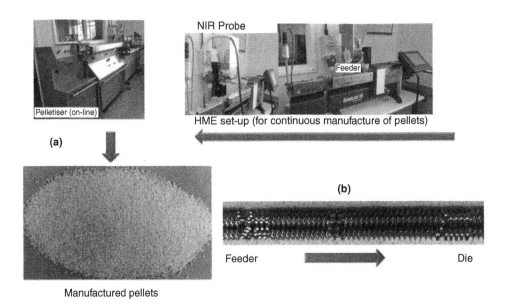

**Figure 7.2** (a) Continuous manufacturing process of SH pellets and (b) screws used for extrusion

### 7.2.2 Advanced Surface Analysis

Surface morphology was examined by SEM for both the drug and extrudates. The extrudates containing both polymers exhibited no drug crystals on the extrudate surface with SH (**Figure 7.3**). A homogenous particle distribution was observed on the surface of all extruded pellets [13]. SEM-EDX was performed to determine the distribution of SH in continuously manufactured pellets. Both of the polymeric carriers used mainly contain C and O atoms, similar to SH. The distribution of SH was visualised by EDX elemental mapping of C and O. The majority of the occupied elements detected in the EDX analysis were carbon, comprising a majority of API and additives for oral drug delivery including SH and polymers, which were employed in this study. This result provided evidence that the pellets were well formed in the intended drug–polymer ratio without the occurrence of uneven content strain during the continuous production of pellets *via* HME processing. It was also evaluated as to whether SH was homogenously dispersed into the particle matrix by means of EDX mapping analysis. Recently, there have been some reports about EDX mapping used for the observation of homogenous distribution of API on extruded formulation(s) (EF).

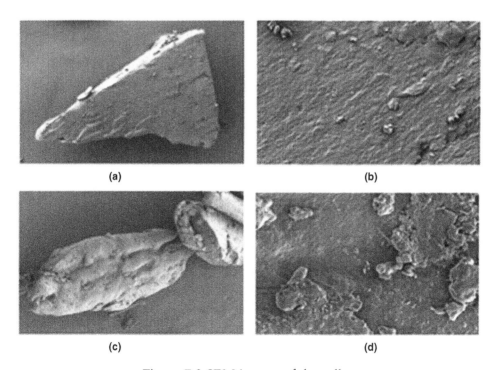

(a)

(b)

(c)

(d)

**Figure 7.3** SEM images of the pellets

### 7.2.4 Thermal Analysis

Differential scanning calorimetry (DSC) was used to analyse the solid-state of pure API, polymers, their PM and active EF. The overall findings from DSC results are summarised in **Figure 7.4**. The DSC scan of pure SH showed an endothermic transition corresponding to its melting point at 226.12 °C ($\delta H$= 114.92 J/g, peak height 17.80 mW) with an onset at 223.82 °C [26]. Similarly, the pure polymers showed glass transition temperature ($T_g$) at 133.01 °C corresponding to $T_g$ of EC N10 and 117.61 °C corresponding to $T_g$ of EC P7, respectively. Even though all binary physical drugs/polymer blends exhibited endothermic peaks (**Figure 7.4**) corresponding to the initial substances at slightly shifted temperatures indicating the drug existence in its crystalline form, the melting peaks were absent in all EF. In the drug/polymer PM, two endothermic peaks were visible (**Figure 7.4**), one at 202.87 °C corresponding to the melting peak of crystalline SH present in the mixture and another one at the lower temperature range at 72.27 °C corresponding to the $T_g$ of EC N10. Similar transitions were observed in the SH/EC P7 system as well where the endothermic peak at 185.86 °C corresponds to the SH melting transition and 61.66 °C to the $T_g$ of EC P7. This huge shift of the melting peak of SH in SH/EC P7 is due to the slight lower $T_g$ value compared to that of EC N10. This also suggests that EC P7 would be more effective than EC N10 for processing SH below the melting temperature. Furthermore, the EF exhibited a broad endothermic peak ranging from 47.83 to 49.19 °C indicating the presence of the drugs in their amorphous forms. It has been observed previously [26] that the position of shifted endothermic peak in the EF is in between the thermal transitions of amorphous drug and polymers.

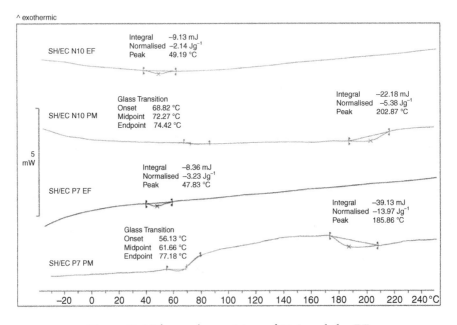

**Figure 7.4** Thermal transition of PM and the EF

This thermal phenomenon complements the formation of a solid dispersion of miscible SH in the amorphous polymer matrices. According to the Gordon Taylor equation, the miscible drug/polymers extrudates would exhibit a broad single $T_g$ at the intermediate position of the $T_g$ of amorphous drug and polymers [24]. Nevertheless, the characteristic peak of SH cannot be found in the heating curve of the EF, indicating that the extruded formulation is different as the SH is present in an amorphous form compared to the PM of drug/polymer.

In order to validate the results obtained from previous DSC, HSM studies were conducted to visually determine the live thermal transitions and the extent of drug melting within the polymer matrices at different stages of heating. SH spread directly onto the glass slide showed minimal change following heating through 180–200 °C (data not shown), which was distinctive in the DSC transition as no thermal transitions did occur until approximately 226 °C. In strong conformity with the DSC data, SH in the EC N10 system showed nominal API dissolution until reaching temperatures of 195–202 °C and thereafter showed extensive solubilisation of the drug. For combinations of SH and EC P7, optical imaging showed that the drug was extensively melted and solubilised in the polymer matrices at temperatures as low as 185 °C, which was also well correlated with the results obtained by DSC testing. Assessment of the solubility parameters as well as atomistic QM based molecular modelling for the SH/EC N10/EC P7 composition also showed minimal difference, indicating similar interaction energies during mixing suggesting that the drug is well miscible with both polymers used. Therefore, solubility parameters and advanced QM based molecular modelling may also provide a good indication for the ability of a drug to be miscible/solubilised within the polymer matrices in order to show better product performances [26].

However, successive production along with the right selection of drug/polymer *via* HME process yielded amorphous solid dispersion as confirmed by DSC. This indicated that consideration of material properties and pre-formulation screening can provide insight into the melt behaviour, but may also be prone to indicate false negatives due to the lack of shear provided by the characterisation techniques at later stage after extrusion.

### 7.2.5 Amorphicity Analysis

The drug–polymer extrudates, including pure drugs and PM of the same composition were studied by X–ray analysis and the diffractograms were recorded to examine the API crystalline state. As depicted in **Figure 7.5** the diffractograms of pure SH presented distinct peaks at 5.75, 14.50, 16.08, 17.42, 18.79, 19.42, 23.22, 29.21, 30.14 2θ values.

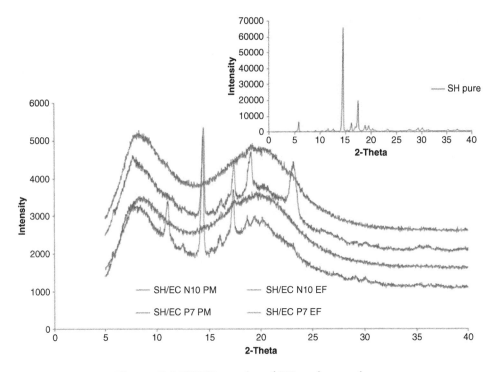

Figure 7.5 XRPD results of EF and pure drug

The PM of both formulations presented identical peaks at lower intensities suggesting that both drugs retain their crystalline properties. No distinct crystalline peaks were found in the EF in EC P7 systems. In contrast almost no distinct intensity peaks (apart from a very low intensity peak at about 15 2θ position) for SH was observed in the diffractogram of the EF in EC N10 system. The absence of SH intensity peaks indicates the presence of amorphous API in the extruded solid dispersion.

### 7.2.6 In Vitro *Dissolution Studies*

Polymeric coating of the compressed tablets with 15% pH dependent polymer proved sufficient to provide 2 h lag time in an acidic medium (**Figure 7.6**). The composition of the matrix core of the final tablet formulation included a combination of hydrophilic (HPMC) and lactose along with extruded EC to control SH release patterns. The combination of all three ingredients provided retarded-release profiles of SH at higher pH values when the dissolution of the coating layer occurred [27]. The release profile was also controlled from the polymer amount coated on the surface of the matrix tablet.

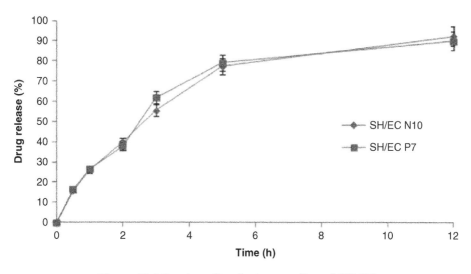

**Figure 7.6** *In vitro* dissolution studies of SH EF

The release mechanism from the tablets can be described as a diffusion process of the drug through the slow release polymers' matrix. The slow dissolving nature of EC polymers hinders the dissolution medium penetration and consequently controls the drug dissolution and diffusion rate [27]. While processing through HME the active substance was covered by EC matrix which resulted in retarded-release of SH over 12 h. Moreover, the polymeric coating level indicated a controlling effect due to its slow dissolution, irrespective of the tablet composition. Lower coating levels (15%) allowed greater penetration of the dissolution medium.

### 7.2.7 Analysis of Release Mechanism

Analysis was performed by the above equations using water as the dissolution test medium and results are summarised in **Table 7.1**. Neither coated nor uncoated tablets fit with **Equation 7.1** of zero order kinetics. Concerning **Equations 7.2–7.6**, regression analyses were performed in the ranges in which linearity was observed. The first order kinetics model plot shows the relationship between the logarithm of the drug residual rate and time on the basis of **Equation 7.2**. Analyses were performed in the ranges in which linearity was maintained with both coated and uncoated tablets. The SH release from both tablets was affected slightly by the coating. In the uncoated tablet, satisfactory linearity was observed throughout the test ($R2 = 0.9773$) for both polymers, and the release pattern was apparently a linear release. In the coated tablets, satisfactory linearity was not observed up to 2 h of dissolution study as no SH was released due to the pH dependant polymer coating. After 2 h, with the increased pH,

SH started releasing and therefore a satisfactory linearity between the logarithm of drug residual rate and time was observed throughout the rest of the test ($R^2 = 0.9657$).

| Table 7.1 Dissolution rate constants and determination coefficients of SH release from coated and uncoated tablets; n dissolution exponent | | | | | |
|---|---|---|---|---|---|
| Dissolution models | | SH/EC N10 granules | SH/EC P7 pellets | SH/EC N10 granules | SH/EC P7 pellets |
| First order | K1 (h-1) | 0.29 ± 0.013 | 0.31 ± 0.027 | 0.4215 ± 0.057 | 0.456 ± 0.068 |
| | R2 | 0.9922 | 0.9709 | 0.9295 | 0.9153 |
| Higuchi | K2 (% hr-1/2) | 29.29 ± 1.56 | 29.48 ± 2.07 | 34.41 ± 2.30 | 34.418 ± 2.303 |
| | R2 | 0.9498 | 0.9113 | 0.9040 | 0.9040 |
| Hixson–Crowell | K3 (% h-1/3) | 0.08 ± 0.005 | 0.089 ± 0.008 | 0.1182 ± 0.176 | 0.126 ± 0.197 |
| | R2 | 0.9838 | 0.9622 | 0.8964 | 0.8829 |
| Krosmer–Peppas | K4 (hr-n) | 31.32 ± 3.91 | 32.95 ± 5.14 | 40.48 ± 4.77 | 42.20 ± 6.11 |
| | R2 | 0.9532 | 0.9214 | 0.9375 | 0.8966 |
| | n | 0.46 ± 0.06 | 0.44 ± 0.08 | 0.40 ± 0.069 | 0.40 ± 0.069 |

The Higuchi model plot shows the relationship between the drug release rate and the square root of time on the basis of **Equation 7.3**. It shows the results of SH release from HME tablets. Analyses were performed in the ranges in which linearity was maintained with both coated and uncoated tablets ($R^2 = 0.9293$–$0.838$). The release from the coated tablet was slightly affected by the coating as expected. However, the release from the uncoated tablet was quite linear throughout the test.

According to the Hixson–Crowell model plot which shows the relationship between the cubic root of the drug residual rate and time on the basis of **Equation 7.5**, Hixson and Crowell showed that the law of cubic root is valid in uniform particles. This equation is based on the assumption that the release occurs only in the vertical direction relative to the matrix surface, and that the release progresses with proportional decreases in all dimensions of the matrix which maintains its shape [17]. Analyses were performed in the ranges in which linearity was maintained with both tablets. The release from uncoated tablets was not linear as the R2 value was found to be very low (0.3854). On the other hand, the SH release from coated tablets according to the Hixson–Crowell model is quite linear throughout the whole test. According to the Hixson–Crowell theory of drug release, it is presumed that the release progression has been maintained, with the decreases in all dimensions of the matrix due to the applied coating on the tablet surfaces and thus maintaining the shapes of the tablets.

**Equation 7.6** is a Korsmeyer–Peppas model equation concerning the diffusion mechanism, and it has been evaluated with reference to pharmaceutical preparations of many matrix types. When both terms of **Equation 7.6** are converted to logarithms, the equation becomes:

$$lnF = ln\ K_4 + n\ ln\ t \tag{7.7}$$

and the slope $n$ can be determined by plotting the logarithm of the release rate against the logarithm of time.

The exponent ($n$) determined by this equation suggests that the uncoated tablets show Fick's diffusion (super case-II transport) when $n = 0.97$. It has been reported that a non-Fickian type release which is also known as anomalous transport occurs only when $0.45 < n < 0.89$, case-II transport can be observed when $n = 0.89$, and super case-II transport when $n > 0.89$ (**Table 7.1**). From the results of analysis of the apparent diffusion pattern, the value of $n$, represented the diffusion pattern as super case transport release pattern [17].

## 7.3 Conclusions

In the current study, HME was used as a robust technique to develop SR extruded tablet of SH *via* a continuous manufacturing platform. Advanced molecular modelling was used as a strong predictive tool to estimate the possible drug-polymer interactions to develop solid dispersions. The tablet was designed to release SH in a pattern that imitates a healthy individual. The development of the SH extruded tablet by HME is preferable since its preparation is quick, economical and with fewer process steps which will make the future pilot scale formulation developments much easier.

## References

1.  J. Breitenbach, *European Journal of Pharmaceutics and Biopharmaceutics*, 2002, **54**, 107.

2.  M.A. Repka, S.K. Battu, S.B. Upadhye, S. Thumma, M.M. Crowley, F. Zhang, C. Martin and J.W. McGinity, *Drug Development and Industrial Pharmacy*, 2007, **33**, 1043.

3.  F. Cilurzo, I. Cupone, P. Minghetti, F. Selmin and L. Montanari, *European Journal of Pharmaceutics and Biopharmaceutics*, 2008, **70**, 895.

4.  F. Zhang and J.W. McGinity, *Pharmaceutical Development and Technology*, 1999, **4**, 241.

5.  F. Kianfar, B. Chowdhry, M. Antonijevic and J. Boateng, *AAPS Journal*, 2011, **13**, 1550.

6.  H.H. Grunhagen and O. Muller, *Pharmaceutical Manufacturing International*, 1995, **1**, 167.

7.  J. Breitkreutz, F. El-Saleh, C. Kiera, P. Kleinebudde and W. Wiedey, *European Journal of Pharmaceutics and Biopharmaceutics*, 2003, **56**, 255.

8.  C. De Brabander, C. Vervaet, L. Fiermans and J.P. Remon, *International Journal of Pharmaceutics*, 2000, **199**, 195.

9.  C. De Brabander, C. Vervaet and J.P. Remon, *Journal of Controlled Release*, 2003, **89**, 235.

10. E. Roblegg, E. Jäger, A. Hodzic, G. Koscher, S. Mohr, A. Zimmer and J. Khinast, *European Journal of Pharmaceutics and Biopharmaceutics*, 2011, **79**, 635.

11. M.R. Clark, T.J. Johnson, R.T. McCabe, J.T. Clark, A. Tuitupou, H. Elgendy, D.R. Friend and P.F. Kiser, *Journal of Pharmaceutical Sciences*, 2012, **101**, 576.

12. K. Vithani, M. Maniruzzaman, S. Mostafa, Y. Cuppok and D. Douroumis in *Proceedings of the 39th Annual Meeting and Exposition of the Controlled Releases Society*, Quebec, Canada, 2012.

13. M. Maniruzzaman, M.A. Hossain and D. Douroumis in *Proceedings of the UK-PharmSci Conference*, Nottingham, UK, 2012.

14.  P.L. Lam, K.K.H. Lee, R.S.M. Wong, G.Y.M. Cheng and S.Y. Cheng, *Bioorganic & Medicinal Chemistry Letters*, 2012, **22**, 3213

15.  B. Abrahamsson, M. Alpsten, B. Bake, U.E. Jonsso, M. Eriksson-Lepkowska and A. Larsson, *Journal of Controlled Release*, 1998, **52**, 301.

16.  S.P. Li, G.N. Mehta, W.N. Grim, J.D. Buehler and R.J. Harwood, *Drug Development and Industrial Pharmacy*, 1988, **14**, 573.

17.  T. Hayashi, H. Kanbea, M. Okada, S. Suzuki, Y. Ikeda, Y. Onuki, T. Kaneko and T. Sonobe, *International Journal of Pharmaceutics*, 2005, **304**, 91.

18.  A.W. Hixson and J.H. Crowell, *Industrial Engineering Chemistry*, 1931, **23**, 923.

19.  T. Higuchi, *Journal of Pharmaceutical Sciences*, 1963, **52**, 1145.

20.  M. Gibaldi and S. Feldman, *Journal of Pharmaceutical Sciences*, 1967, **56**, 1238.

21.  R.W. Korsmeyer, R. Gurny, E.M. Doelker, P. Buri and N.A. Peppas, *International Journal of Pharmaceutics*, 1983, **15**, 25.

22.  P. Costa and J.M. Sousa Lobo, *European Journal of Pharmaceutics and Biopharmaceutics*, 2001, **13**, 123.

23.  M. Maniruzzaman, M.M. Rana, J.S. Boateng, J.C. Mitchell and D. Douroumis, *Drug Development and Industrial Pharmacy*, 2012, **39**, 2, 218.

24.  X. Zheng, R. Yang, X. Tang and L. Zheng, *Drug Development and Industrial Pharmacy*, 2007, **33**, 791.

25.  D.J. Greenhalgh, W. Peter and T.P. York, *Journal of Pharmaceutical Sciences*, 1999, **88**, 1182.

26.  J.C. DiNunzio, C. Brough, J.R. Hughey, D.A. Miller, R.O. Williams, III., and J.W. McGinity, *European Journal of Pharmaceutics and Biopharmaceutics*, 2010, **74**, 340.

27.  M. Maniruzzaman, M.A. Hossain, A. Abiodun and D. Douroumis in *Proceedings of the 39th Annual Meeting and Exposition of the Controlled Releases Society*, Quebec, Canada, 2012.

# 8 Novel Pharmaceutical Formulations Using Hot-Melt Extrusion Processing as a Continuous Manufacturing Technique

Mohammed Maniruzzaman

## 8 Introduction

With the advances of high throughput screening and combinational chemistry, an increasing number of pharmaceutical active compounds have been identified and subsequently developed [1]. More than 40% of newly synthesised compounds are highly hydrophopic and water-insoluble and thus it remains a key challenge for pharmaceutical scientists to improve the aqueous solubility of these poorly water-soluble drugs [1, 2]. Therefore, advanced formulation optimisation and pharmaceutical processing technologies are required to overcome solubility issues and thus enable the oral delivery of poorly water-soluble drugs. A number of approaches have been utilised and investigated for dissolution enhancement including particle size reduction, salt formation, lipid-based formulation and solubilisation [3]. Similarly, various physical techniques have been used to enhance dissolution of various insoluble drugs such as co-evaporation [4], hot spin mixing [5], roll-mixing or co-milling [6], freeze-drying [7], spray drying [8, 9], supercritical fluid processing [10] and hot-melt extrusion (HME) [1, 2, 11]. HME offers several advantages over the aforementioned technologies as it is cost efficient, solvent free, easy to scale-up and it can be developed as a continuous manufacturing process.

HME has been used extensively to enhance the dissolution rate of various poorly water-soluble drug candidates quite often leading to the formation of amorphous or molecularly dispersed drug forms [1, 2] either in polymeric or lipidic carriers. A novel approach involves the co-extrusion of polymer/lipid carriers that could potentially enhance dissolution rates of poorly water-soluble drugs with the aim of providing a synergistic effect. The co-extrusion (Co-HME) approach is defined as a process where melt extrusion is carried out simultaneously in the presence of two or more carriers (matrices) along with the active substance [12]. Hydrophilic polymers have shown a significant increase of drug dissolution rates while several lipids are used to increase solubility of insoluble drugs.

The aim of this study was to investigate the effect of polymer/lipid combinations on the dissolution rates of the water-insoluble drug, Drug X, after Co-HME processing.

## 8.1 Materials and Methods

### 8.1.1 Materials

Drug X was purchased from Sigma Aldrich (UK). Hydrophillic polymer hydroxypropyl methylcellulose (HPMC) and stearoyl macrogol-32 glycerides [G-50, were provided by ShinEtsu (Japan) and Gattefosse (France)], respectively. The high-performance liquid chromatography (HPLC) solvents were of analytical grade and purchased from Fisher Chemicals (UK). All materials were used as received.

### 8.1.2 Hot-Melt Extrusion Processing

The Drug X/G-50/HPMC dry powders were mixed properly in 100 g batches for 10 min. A Turbula TF2 mixer (Switzerland) was used to blend the powder properly for 10 min to achieve uniformity of powder. Formulations were prepared at 30/35/35 drug/polymer/lipid (w/w) ratios. The extrudates (EXT) were processed with a twin-screw extruder (Eurolab 16) in order to obtain polymer/lipid pellets (1 mm) at 50 rpm using 1 kg/h feed-rate. The extrusion process was performed at 50/130/135/135/135/135/135/135/135/135 °C (from feeder to die).

### 8.1.3 Differential Scanning Calorimetry Analysis

The physical state of pure drug, pure lipid, physical mixtures (PM) and EXT was examined by using Mettler-Toledo 823e (Mettler-Toledo, Switzerland). Samples were prepared with an initial weight of approximately 4–5 mg in a sealed aluminium pan with lid. The samples were heated from -30 to 200 °C at 10 °C/min scanning rate under nitrogen ($N_2$) atmosphere. The G-50 was heated from 0 to 80 °C at the same scanning rate due to the low melting point.

### 8.1.4 Hot Stage Microscopy Analysis

The characterisation of Drug X in the molten lipid/polymeric carrier was assessed using hot stage microscopy (HSM). During testing, an Olympus BX60 microscope

(Olympus Corp., USA) with Insight QE camera (Diagnostic Instruments, Inc., USA) was used to visually observe samples, whilst a FP82HT hot stage controlled by a FP 90 central processor (Mettler Toledo, USA) was used to maintain temperatures between 20 and 150 °C. Images were captured under visible and polarised light using Spot Advance Software (Diagnostic Instruments, Inc., USA).

### 8.1.5 X-ray Powder Diffraction

X-ray diffraction was used to assess the solid-state of the EXT. X-ray powder diffraction was performed for bulk drug, lipid, and PM and extruded formulations. A D8 Advance (Bruker, Germany) in theta-theta mode, was used with Cu anode at 40 kV and 40 mA, parallel beam Goebel mirror, LYNXEYE™ Position Sensitive Detector with 3° opening and LynxIris at 6.5 mm, 0.2 mm exit slit and sample rotation of 15 rpm. The samples were scanned from 2° to 40° 2-theta with a step size of 0.02° 2-theta using 0.2 s per step counting time; 176 channels active on particle size distribution making a total counting time of 35.2 per step.

### 8.1.6 In Vitro *Drug Release Studies*

An *in vitro* dissolution study was carried out by using a Varian 705 DS dissolution paddle apparatus (Varian Inc., USA) at 100 rpm and 37 ± 0.5 °C. 100 mg equivalent of Drug X was used in each dissolution vessel. 750 ml of 0.1 M HCl solution was used as the dissolution media to keep the pH 1.2 for first 2 h and then the pH was adjusted to 6.8 using a solution of 0.2 M of dihydrogen sodium orthophosphate (adjusted with NaOH). About 2–3 ml of solutions were withdrawn at predetermined time intervals to determine the release of indomethacin (INM). All dissolutions were performed in triplicate.

### 8.1.7 High-performance Liquid Chromatography Analysis

The release of Drug X was determined by HPLC. An Agilent Technologies system equipped with a HYCROME 4889, 5 µm × 150 mm × 4 mm column at 276 nm was used for the Drug X HPLC assay. The mobile phase consisted of methanol/water/acetic acid (60:40:1, v/v/v). The flow rate was 1.5 ml/min and the retention time for Drug X was about 4 min. The Drug X calibration curves ($R^2 = 0.9998$), at concentrations varying from 10 to 50 µg/ml, were used to evaluate all the samples with 20 µl injection volume.

## 8.2 Results and Discussion

### *8.2.1 Hot-Melt Extrusion Processing*

Extrusion processing was optimised to obtain the EXT as yellowish pellets of 1 mm length (**Figure 8.1**). The temperature, screw speed and the feed-rate (data not shown) were found to be the critical processing parameters for the development of the extruded formulations.

All formulations were easily extruded to produce translucent strands with high INM loadings at 30% (w/w ratio) and high throughput (1 kg/h). All materials were found to be miscible according to the Hansen solubility parameters ($\delta 2$) [1, 13]. The estimated solubility values of INM (22.8 MPa$^{1/2}$), HPMC (25.30 MPa$^{1/2}$) and G-50 (19.8 MPa$^{1/2}$) showed a difference of $\Delta\delta < 7.0$ MPa$^{1/2}$ suggesting complete miscibility with each other. The polymer/lipid formulations were extruded at relative low temperatures due to the addition of the lipid which acted as a plasticiser.

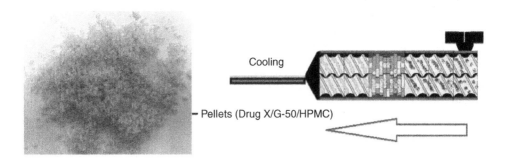

Figure 8.1 Manufacturing the of Drug X pellets by HME processing. AQOAT: HPMC grade polymer from ShinEtsu, Japan

### *8.2.2 Thermal Analysis*

Differential scanning calorimetry (DSC) was used to determine the solid-state of the drug in the extruded matrices as well as to outline any possible intermolecular drug-lipid-polymer interactions. **Figure 8.2** shows the thermal transitions of pure Drug X, G-50 and HPMC. The bulk Drug X showed an endothermic thermal transition at 161.18 °C ($\Delta H$ = 115.33 j/g), which corresponds to its melting peak, while G-50 exhibited an endothermic sharp melting peak at 45.32 °C ($\Delta H$ = 143.67 °C). Similarly, the bulk HPMC pure showed a step change due to glass transition temperature at

128.21 °C. The DSC analysis (**Figure 8.2**) of the extruded materials showed a shift of the hydroxypropyl methylcellulose acetate succinate (HPMCAS)-LF glass transition at the lower temperature of 88.31 °C, while the Drug X melting endotherm disappeared at the 161 °C compared to the bulk substance. These results indicate the existence of Drug X amorphous state within the extruded matrices [1] and it is attributed to the possible drug–lipid–polymer intermolecular interactions during the extrusions.

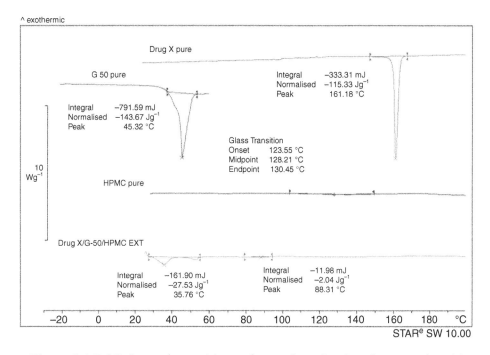

**Figure 8.2** DSC thermal transitions of pure drug, lipid, polymer and EXT

HSM studies were conducted to visually determine the thermal transitions and the extent of drug melting within the polymer/lipid matrices at different stages of heating. Images taken using HSM under optical light are shown in **Figure 8.3**. The bulk Drug X showed no changes up to 150–160 °C, which is in agreement with the DSC results, as thermal transition due to the melting of the drug occurred at 161 °C. Similar to the DSC thermograms, the Drug X/G-50/HPMC EXT showed nominal API melting until temperatures of 100–120 °C were reached and thereafter showed extensive melting of the drug (**Figure 8.3**). This could potentially be attributed to the intermolecular interactions or possibly the solubilisation of the drug into polymer/lipid matrices during the extrusion processing.

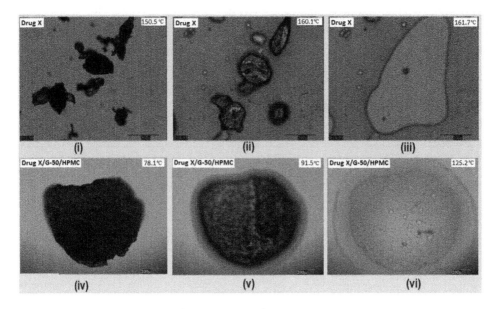

Figure 8.3 HSM themograms of Drug X/G-50/HPMC EXT

### *8.2.3 X-ray Powder Diffraction Analysis*

The drug–polymer–lipid EXT, including pure drug and PM of the same composition were studied by X-ray analysis and the diffractograms were recorded to examine the crystalline state of Drug X. As can be seen from **Figure 8.4a,** the diffractogram of pure INM showed distinct intensity peaks at $10.17°$, $11.62°$, $17.02°$, $19.60°$, $21.82°$, $23.99°$, $26.61°$, $29.37°$, $30.32°$, $33.55°$ degree $2\theta$. The PM of Drug X formulation showed identical peaks at lower intensities suggesting that the drug retained its crystallinity. In contrast, no distinct intense peaks due to Drug X were observed in the diffractograms of the extruded formulation. The absence of Drug X intensity peaks indicates the presence of amorphous Drug X in the polymer/lipid matrix complementing the findings from DSC. However, Drug X/ HPMC (binary) extruded formulations revealed distinct intensity peaks of the drug at $12–17°$ $2\theta$ and $22–27°$ $2\theta$ positions (**Figure 8.4b**), simply indicating the crystalline existence of the drug in the extruded formulation. This could be due adverse or incomplete mixing during the extrusion processing, while the presence of the lipid as the third excipient has facilitated the homogenous mixing and thus the interactions. As a result, due to the presence of a lipidic carrier as a third excipient in the formulation, a formation of the amorphous solid dispersions has been developed and optimised.

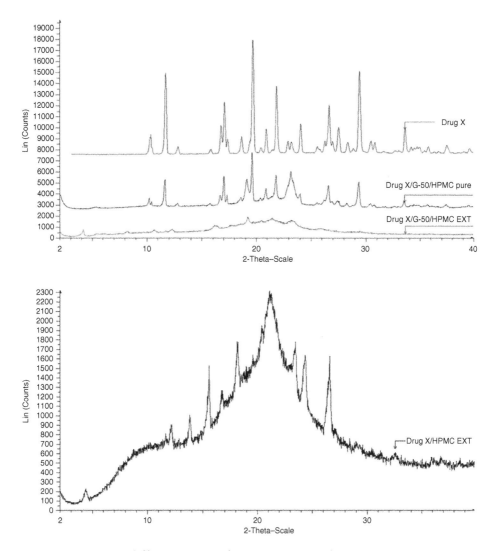

**Figure 8.4** a) X-ray diffractograms of Drug X pure and Drug X/G-50/HPMC PM and Drug X/G-50/HPMC EXT and b) X-ray diffractograms of Drug X/HPMC EXT

### *8.2.4* In vitro *Dissolution Studies*

The dissolution behaviour of the processed formulations was assessed for extruded granules (average particle size 250 μm) and pellets (1 mm) in comparison to the bulk Drug X powder.

As shown in **Figure 8.5**, a lag time with no drug release was observed for 2 h in acidic dissolution media. This was attributed to the pH dependency of HPMC which dissolves at a higher pH of 5.5. It appears that the polymer has a predominant effect and prevents drug release at low pH values. At higher pH values, Drug X was rapidly released within 1 h and no significant difference ($f_2$ test) was observed between the extruded pellets and granules. Interestingly, HPMC/G-50 showed a synergistic effect resulting in faster release rates for Drug X compared to those of Drug X/HPMC EXT. It can be clearly seen that co-extruded polymer/lipid formulations significantly affected INM dissolution rates (**Figure 8.5**). The synergistic effect of G-50/HPMC revealed a release of about 90% in 30 min while for HPMC/Drug X only 20% was released after 3 h in basic media.

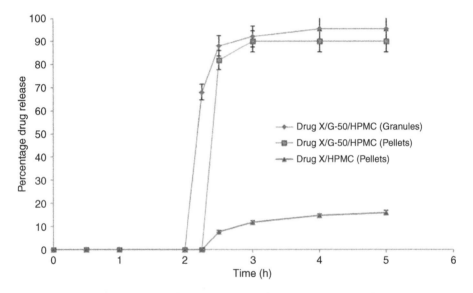

Figure 8.5 *In vitro* release profiles of Drug X from extruded pellets, granules (pH 1.2 for 1st 2 h and then pH 6.8 for rest 3 h; n = 3, 37 °C, paddle speed 100 rpm)

## 8.3 Conclusions

In this study novel polymer/lipid formulations were processed *via* HME to increase the dissolution rates of the poorly water-soluble Drug X. The optimised process resulted in high Drug X loaded formulations in the form of pellets or granules. At low pH values (acidic media) the HPMC grade polymer presented a predominant effect and no drug release was observed for 2 h (lag time). In comparison, high Drug X dissolution rates were observed in alkaline media where the polymer/lipid formulations showed a synergistic effect.

## References

1.  M. Maniruzzaman, M. Rana, J.S. Boateng and D. Douroumis, *Drug Development and Industrial Pharmacy*, 2013, **392**, 218.

2.  A. Grycze, G.S. Schminke, M. Maniruzzaman, J. Beck and D. Douroumis, *Colloids and Surfaces B: Biointerfaces*, 2011, **86**, 275.

3.  L. Dierickx, B. Van Snick, T. Monteyne, T. De Beer, J.P. Remon and C. Vervaet, *European Journal of Pharmaceutics and Biopharmaceutics*, 2014, **88**, 502.

4.  S.W. Hong, B.S. Lee, S.J. Park, H.R. Jeon, K.Y. Moon, M.H. Kang, S.H. Park, S.U. Choi, W.H. Song, J. Lee and Y.W. Choi, *Archives of Pharmaceutical Research*, 2011, **34**, 127.

5.  M. Dittgen, S. Fricke, H. Gerecke and H. Osterwald, *Die Pharmazie*, 1995, **50**, 225.

6.  J. Breitenbach, *European Journal of Pharmaceutics and Biopharmaceutics*, 2002, **54**, 107.

7.  H. Sekikaw, W. Fukuda, M. Takada, K. Ohtani, T. Arita and M. Nakano, *Chemical and Pharmaceutical Bulletin*, 1983, **31**, 1350.

8.  V. Caron, L. Tajber, O.I. Corrigan and A.M. Healy, *Molecular Pharmaceutics*, 2011, **8**, 532.

9.  K. Wu, J. Li, W. Wang and D.A. Winstead, *Journal of Pharmaceutical Sciences*, 2009, **98**, 2422.

10. K. Gong, R. Viboonkiat, I.U. Rehman, G. Buckton and J.A. Darr, *Journal of Pharmaceutical Sciences*, 2005, **94**, 2583.

11. M. Maniruzzaman, J.S. Boateng, M.J. Snowden and D. Douroumis, *International Scholarly Research Notices: Pharmaceutics*, 2012, Article ID:436763.

12. A-K. Vynckiera, L. Dierickxa, J. Voorspoelsb, Y. Gonnissenb, J.P. Remon and C. Vervaeta, *Journal of Pharmacy and Pharmacology*, 2013, **66**, 167.

13. R. Witzleb, V.R. Kanikanti, H.J. Hamann and P. Kleinebudde, *European Journal of Pharmaceutics and Biopharmaceutics*, 2011, **77**, 170.

14. A.M. Chuah, B. Jacob, Z. Jie, S. Ramesh, S. Mandal, J.K. Puthan, P. Deshpande, V.V. Vaidyanathan, R.W. Gelling, G. Patel, T. Das and S. Shreeram, *Food Chemistry*, 2014, **156**, 227.

# 9 Continuous Polymorphic Transformations Study *via* Hot-Melt Extrusion Process

Mohammed Maniruzzaman

## 9 Background

Different crystal modifications (known as polymorphs) of the same molecule exhibit variations in physical properties such as solubility, stability, optical properties and melting temperatures ($T_m$). These can cause far reaching ramifications on their production and applications. The term 'polymorphism' is considered an important factor with particular importance to the pharmaceutical industry since the solid-state of an active pharmaceutical ingredient (API) can drastically alter drug bioavailability. Preparative conditions such as surface properties and trace quantities of impurities as well as the temperature and pressure, can control the crystallisation of a particular solid-state.

## 9.1 Polymorphism

Polymorphism is referred to as the ability of a crystalline substance to exist in two or more crystalline forms. This term is an important aspect of crystallisation products and processes. The fact that chemical species can exist in different lattice structures often leads to significant variations in the physical properties of such crystals such as thermodynamic, spectroscopic, interfacial and mechanical properties. Solubility and shape, as well as compression and filtration properties, dissolution rate and bioavailability, can get altered with the transformation of the polymorphic forms. A slight modification in the crystallisation process, such as a variation of the cooling rate or of the agitation, can produce a different polymorph.

Solid solutions are analogous to liquid solutions which consist of just one phase irrespective of the number of components. Sekiguchi and co-workers (1960) [1] were the first to report a solid solution of two components (which is now known as a eutectic mixture), being completely miscible in the liquid state but only miscible to a limited extent in the solid-state. To date, the preparation and optimisation of solid dispersions of active drug substances is still a prime focus in pharmaceutical formulation development and research. Various techniques have been reported for the preparation of solid dispersions such as hot spin mixing [2]; spray drying [3, 4]; co-evaporation or co-precipitation [5]; freeze-drying [6]; supercritical fluid processing [7]; and hot-melt extrusion (HME) [8, 9]. In recent years, however, the HME method has, arguably, undergone a renaissance in relation to formulation development since extrusion of moistened powders has been well known for many years in the processing of pharmaceutical formulations.

HME is a contemporary method for the manufacture of solid dispersions that is becoming more widely utilised in the production of drug delivery systems; it combines the advantages of a solvent-free solid dispersion with that of a dust-free processing environment. However, during and after HME processing it is necessary to obtain as detailed an understanding as possible of the physical state of the drug(s) in the polymeric matrices used as drug carriers. The main reason for the foregoing is that the solid-state properties of the drug can affect both the stability and the dissolution behaviour of the pharmaceutical formulation obtained. In terms of solid dispersions, it is well known that the API may exist in a number of physical states in polymer matrices [10, 11]. These include crystalline dispersions (substitutional and interstitial), molecular dispersions (continuous or discontinuous) and amorphous dispersions (the drug is present as a separate amorphous phase). Previous studies e.g., reference [11] have demonstrated that the molecular nature of solid-state dispersions and/or phase separation in melt extruded formulations (EF) can lead to crystallisation or polymorphic conversions of the API during the extrusion process. However, identification, prediction and characterisation of solid-state dispersions involving polymorphic conversions are difficult. One reason being that, to date, the relationship between solid-state dispersions and their suitability as e.g., drug delivery systems is neither well understood nor predictable, while being acknowledged to be of considerable importance.

## 9.2 Case Study: Polymorphic Transformation of Paracetamol

Paracetamol (PMOL) is a white crystalline powder mainly used as an analgesic and antipyretic. The rationale underlying the selection of PMOL as a model drug for this study is, primarily, its existence in multi-polymorphic forms such as Form I (monoclinic), Form II (orthorhombic) and Form III (metastable), in decreasing order of stability and melting point [9]. This provides an interesting system for studying the effects of processing and conversion of crystal forms of drugs. Furthermore, amorphous PMOL exhibits low stability as the glass transition occurs at about 25 °C [9].

Variable temperature X-ray powder diffraction (VTXRPD) analysis has, to a limited extent, previously been used to characterise solid-state pharmaceutical reactions including crystal transformations [12, 13]. This technique is a powerful tool that can be used to explore such changes since it permits simultaneous quantification of multiple solid phases [14, 15].

In the current study, the effect of temperature on the transformation of PMOL crystal Form I in physical mixtures (PM) of drug/copolymers and hot-melt extruded solid dispersions is exploited by using thermal analysis, VTXRPD as well as in-line near-infrared (NIR) spectroscopy. The aforementioned approach could, potentially, be employed as an effective tool to predict and characterise polymorphic transformations in continuous pharmaceutical manufacturing processing when using HME.

### 9.2.1 Experimental Methods

PMOL was kindly donated by Mallinckrodt Chemical Ltd (UK). Soluplus® (SOL) and Plasdone® S630 [N-vinylpyrrolidone (VP)/vinyl acetate (VA)] were also donated from BASF Polymers (Germany) and ISP (Germany), respectively.

### 9.2.2 Theoretical Calculation

The Hansen Solubility Parameter [16] was used to predict the miscibility of drugs with polymers in solid dispersions. The Hansen Solubility Parameters ($\delta$) of both drugs as well as the polymers were calculated by considering their chemical structural orientations. In order to determine the theoretical drug–polymer miscibility the solubility parameters were calculated by using the Van Krevelen–Hoftyzer method [17] according to **Equation 9.1**:

$$\delta^2 = \delta_d^2 + \delta_p^2 + \delta_h^2 \tag{9.1}$$

where,

$$\delta_d = \frac{\sum F_{di}}{vi}, \delta_p = \frac{\sqrt{\sum F_{pi}^2}}{vi}, \delta_h = \sqrt{\frac{\sum E_{hi}}{vi}}$$

$i$ = structural groups within the molecule, $\delta$ = the total solubility parameter. $F_{di}$ = molar attraction constant due to molar dispersion forces, $F_{pi}^2$ = molar attraction constant due to molar polarisation forces, $E_{hi}$ = hydrogen bonding energy, and $V_i$ = group contribution to molar volume. The F–H interaction parameter, $\chi$, of the model system was determined at two different conditions using the Nishi–Wang **Equation 9.2** [18] equation based on melting point depression data and Hildebrand—Scott **Equation 9.3** [19] correlations with solubility parameter. The F–H interaction parameter ($\chi$) for all of the drug/polymers binary mixtures were calculated by using **Equations 9.2** and **9.3**. The value determined by **Equation 9.2** represents the interactions between the two substances, specifically at the $T_m$, which may not be extrapolated to other temperatures:

$$\frac{1}{T_m} - \frac{1}{T_m^0} = -\frac{R\upsilon_{drug}}{\Delta H_{drug}\upsilon_{poly}}[\ln\phi_{drug} + (1 - \frac{1}{m_{poly}}) \times (1 - \phi_{drug}) + \chi_{drug-poly}(1 - \phi_{drug})^2] \quad (9.2)$$

where, $\upsilon$ is the molar volume of the repeating unit, $m$ is the degree of polymerisation, $\phi$ is the volume fraction and $\chi$ is the crystalline–amorphous polymer interaction parameter, $T_m$ and $T^0_m$ is the crystalline melting peak and amorphous glass transition temperature ($T_g$) in the system, respectively. F–H interaction parameter ($\chi$) can be also estimated by the method developed by Hildebrand and Scott according to **Equation 9.3** [19]:

$$\chi = \frac{\upsilon(\delta_{drug} - \delta_{poly})^2}{RT} \quad (9.3)$$

where $R$ is the gas constant, $T$ is the absolute temperature, and $\upsilon$ the volume per lattice site and $\delta_{drug}$ and $\delta_{poly}$ are solubility parameters of drugs and polymers respectively.

### 9.2.3 Continuous Hot-Melt Extrusion Process and In-line Monitoring

PMOL formulations with both SOL and VP/VA to be extruded, were mixed properly in 100 g batches for 10 min each prior to the extrusion. A Turbula TF2 Mixer was used to blend the powder formulations thoroughly. The drug/polymer ratio used was 40:60, 50:50 and 60:40 for both polymers. Extrusion of all PMOL based formulations was performed using a Eurolab 16 twin-screw extruder (Thermo Fisher, Germany) equipped with a 2 mm rod die and with a screw speed of 50–100 rpm (feed-rate 0.5–1 kg/h). The temperature profile used for all formulations was 50/100/115/120/120/120 °C (from feeding zone → die). The produced extrudates (EXT) (strands) were milled for 5 min at 400 rpm by a Pulverisette ball milling system (Retsch, Germany) to obtain granules (<500 μm). NIR spectrometry was performed during extrusion using an Antaris II NIR spectrometer (Thermo Scientific, UK) equipped with a halogen NIR source. During the process an InGaAs detector was employed for in-line monitoring. The experimental apparatus contained a fibre-optic probe which was connected to the NIR spectrometer. NIR spectra were collected in real-time during the entire extrusion process *via* a fibre-optic diffuse reflectance probe. All NIR in-line spectra were continuously collected using the results from 3.0 integration software (Thermo Fisher Scientific, Germany). Each spectrum was acquired by averaging 32 scans with a resolution of 16 cm$^{-1}$ and over the range of 4,000–10,000 cm$^{-1}$. The acquisition of a full spectrum took approximately 6 s. The spectral pre-processing was performed using the TQ analyst 8.6.12 software (Thermo Fisher Scientific, Germany).

### 9.2.4 Thermal Analysis

A Mettler-Toledo 823e (Greifensee, Switzerland) differential scanning calorimeter was used to carry out differential scanning calorimetry (DSC) runs of pure actives, PM and EXT. 3–5 mg of sample was placed in sealed aluminium pans with pierced lids. The samples were heated at 1–10 °C/min heating rate and from 0 to 220 °C under dry nitrogen atmosphere. In addition, modulated temperature DSC studies were performed from 20 to 150 °C temperature range with an underlying heating rate of 1 °C/min to further analyse the samples. The pulse height was adjusted to 1–2 °C with a temperature pulse width of 15–30 s.

Characterisation of PMOL in the molten polymeric carrier was assessed using hot stage microscopy (HSM). During testing, an Olympus BX60 microscope (Olympus Corp., USA) with Insight QE camera (Diagnostic Instruments, Inc., USA) was used to visually observe samples, while a FP82HT hot stage controlled by a FP 90 central processor (Mettler Toledo, USA) maintained temperatures at 20–250 °C. Images were captured under visible and polarised light using Spot Advance Software (Diagnostic Instruments, Inc., USA).

VTXRPD was used to determine the polymorphs with variable temperatures of pure active substance, PM and extruded materials using a D8 Advance (Bruker, Germany) in theta-theta reflection mode. For the purpose of the study, a copper anode at 40 kV and 40 mA, parallel beam Goebel mirror, 0.2 mm exit slit, LYNXEYE silicon strip Position Sensitive Detector (PSD) opening at 3° with 176 active channels, LynxIris at 6.5 mm, secondary 2.5° Soller slit were used. Each sample was scanned from 2 to 40° 2θ with a step size of 0.02° 2θ and a counting time of 0.2 seconds per step; 176 channels active on the PSD making a total counting time of 35.2 seconds per step. Variable temperature was achieved by using Anton Paar TTK450 non-ambient sample chamber with maximum heating rate 0.2 °C per second. Data collection was performed with DiffracPlus Commander while the data manipulation and presentation were carried out by using EVA V.16 software [20].

## 9.3 Results and Discussion

### 9.3.1 Continuous Extrusion Process and Theoretical Consideration

The PMOL miscibility with SOL and VP/VA was estimated prior to extrusion by determining the Hansen Solubility Parameter using the Van Krevelen–Hoftyzer approach for pure PMOL and both polymers. The miscibility is caused by balancing the energy of mixing released by intermolecular interactions between the components and the energy released by intramolecular interactions within the components [21, 22]. The theoretical approach of the solubility parameter suggests that compounds with similar $\delta$ values are likely to be miscible. It is usually accepted that compounds with $\Delta dt < 7$ MPa$^{1/2}$ are likely to be miscible and compounds with $\Delta dt > 10$ MPa$^{1/2}$ are likely to be immiscible [21]. Thus, solubility parameters provide a simple and generic capability for rational selection of carriers in the preparation of solid dispersions. As it can be seen in **Table 9.1**, the difference between the calculated solubility parameters of the polymers and the drug indicate that PMOL is likely to be miscible with both SOL and VP/VA. The possible solubility/miscibility of PMOL in VP/VA is higher than that of SOL, as the difference of the solubility parameter between PMOL and VP/VA is only 3.99 MPa$^{1/2}$, while for PMOL/SOL it is 6.34 MPa$^{1/2}$.

| Table 9.1 Calculated Hansen Solubility Parameters for PMOL and copolymers | | | | | |
|---|---|---|---|---|---|
| Sample | $\delta_d$ (MP$_a{}^{1/2}$) | $\delta_p$ (MP$_a{}^{1/2}$) | $\delta_v$ (MP$_a{}^{1/2}$) | $\delta_h$ (MP$_a{}^{1/2}$) | $\delta_t$ (MP$_a{}^{1/2}$) | $\Delta\delta$ |
| PMOL | 19.43 | 9.71 | 29.14 | 13.88 | 25.77 | – |
| SOL | 15.14 | 0.45 | 15.15 | 12.18 | 19.43 | 6.34 |
| VA64 | 18.0 | 0.64 | 18.01 | 7.73 | 19.60 | 6.17 |
| VA64: Kollidon® VA 64 | | | | | | |

In order to determine the F–H interaction parameter, $\chi$ [23, 24] between PMOL and both polymers, the heat of fusion of crystalline PMOL as well as the $T_g$ of both polymers and melting peaks of API (**Table 9.2**) were determined by DSC scans. The molar volumes of two different polymers SOL and VP/VA as well as PMOL, were estimated from the functional group contribution of the chemical structures. The molecular volume calculated for PMOL is 120.95 cm$^3$ while the molecular volumes for SOL and VP/VA are 380.04 cm$^3$ (monomer) and 206.33 cm$^3$ (monomer), respectively. From **Equations 9.3** and **9.4** the average value of $\chi$ is calculated as shown in **Table 9.2**. The accurate estimation of the interaction parameter by using F–H theory was straightforward as $\chi$ depends on multiple factors such as crystalline $T_m$ of API, $T_g$ of polymers, molecular volumes and degree of polymerisations.

| Table 9.2 F–H interaction parameters for melt extruded drug–copolymer formulations | | | |
|---|---|---|---|
| Formulation (w/w%) | Volume fractions (drug/copolymer) (ø) | Nishi–Wang (-$\chi$) | Hildebrand–Scott ($\chi$) × $10^{-3}$ |
| PMOL/SOL (40%) | 0.4:0.6 | 4.3 ± 0.01 | 1.6 |
| PMOL/SOL (50%) | 0.5:0.5 | 4.9 ± 0.01 | |
| PMOL/SOL (60%) | 0.6:0.4 | 6.0 ± 0.01 | |
| PMOL/VA64 (40%) | 0.4:0.6 | 4.34 ± 0.01 | 1.5 |
| PMOL/VA64 (50%) | 0.5:0.5 | 4.96 ± 0.01 | |
| PMOL/VA64 (60%) | 0.6:0.4 | 5.98 ± 0.01 | |

Generally, the negative value of a calculated interaction parameter indicates a net attraction force between two molecular compounds in a binary mixture which is favourable for all compositions at observed $T_m$ of PMOL [25, 26]. Therefore lower positive values of $\chi$ suggest stronger interactions between drug/polymers powders in the $T_m$ (both drug and polymers). In **Table 9.2**, it can be seen that VP/VA facilitates stronger interactions than that of SOL with PMOL, which is in a good agreement with the predicted drug/polymer miscibility estimated by the Hansen Solubility Parameter. Interestingly, by increasing the amount of PMOL in the SOL based formulations the interaction strength decreases. This could be attributed to the presence of higher number of –NH groups and their contribution towards the interaction parameter compared to that of –COOH groups. The lower number of groups contributions corresponding to the –COOH molecules reduces the net attraction forces between drug and polymer. It has previously been reported that interaction can take place between the amine group of the drug and the carboxyl group of the polymer [8, 9]. Quite similar trends have been observed for PMOL-VP/VA formulations with ascending trends of the interaction parameter values with increasing drug contents in the formulations. The calculated $\chi$ values indicated that the higher the positive values the less the attraction forces, thus weak or no drug/polymer interactions [25]. These findings along with the data obtained from Hansen Solubility Parameter complement each other, showing better drug/polymer miscibility in VP/VA systems compared to that of SOL systems.

As a result, we anticipated drug/polymer interactions during the extrusion processing. However, both theoretical approaches cannot predict possible polymorphic transformations in the melting state, apart from indicating possible drug/polymer interaction.

### 9.3.2 Physicochemical Characterisation of the Polymorphic Transformation during Hot-Melt Extrusion

Scanning electron microscopy (SEM) was used to examine the surface morphology of all EF as depicted in **Figure 9.1**. The EF containing PMOL exhibited drug crystals on the EXT surface after extrusion with both polymers. SEM scans of pure PMOL showed a monoclinic structure which indicates the polymorphic Form I of PMOL used (data not shown). Later on, the findings from the SEM analysis was confirmed by X-ray powder diffraction (XRPD) studies where the diffractogram obtained from PMOL bulk powder showed a similar pattern to the monoclinic crystalline structure (found from the X-ray diffraction database, PDF-2 released 2008) [20]. The SEM results as depicted in **Figure 9.1** showed that PMOL is present as an octahedral shape in both SOL and VP/VA polymer systems in all drug/polymer (40–60% wt/wt) ratios. The findings from SEM simply indicate the existence of crystalline PMOL in its polymorphic Form I (monoclinic). The average particle size ranges from 50–250 μm in all the EF.

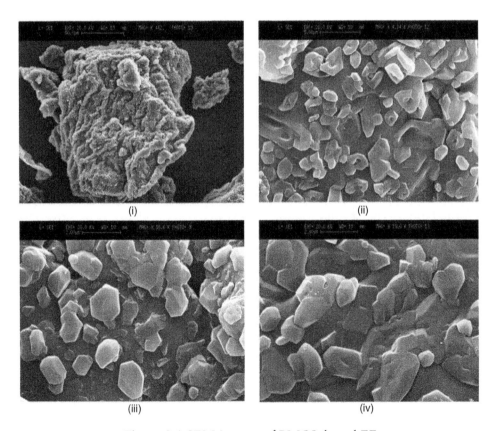

Figure 9.1 SEM images of PMOL based EF

DSC studies were performed to investigate the physical state of the drug within the polymer matrix [27]. The DSC thermogram of pure (purity ~99.96%) PMOL showed a sharp melting peak at 169.10 °C (fusion enthalpy 137.06 j/g) with an onset of peak at 168 °C, which corresponds to the polymorphic Form I [9]. In the case of SOL, an endothermic transition was obtained at 68.4 °C (onset 66.4 °C), whilst for VP/VA an endothermic transition was obtained at 105.5 °C (onset 101.39 °C); both endothermic transitions correspond to the $T_g$ of the polymers.

The DSC scans of the PMOL/SOL EXT (**Table 9.3**) showed melting endotherms at 141.25, 144.6 and 150.8 °C respectively that correspond to 40, 50 and 60% PMOL loadings. The observed melting peaks are shifted to lower temperatures and the peak shapes are broader compared to those of pure PMOL, suggesting the presence of crystalline PMOL. In this case, the observed melting peak of PMOL between 141–151 °C indicates the presence of Form II PMOL in the polymer matrix [9–11]. However, the shifts of the melting endothermic peaks can also be attributed to possible drug/polymer interactions and thus changes in the modifications of the crystal structure.

| Table 9.3 DSC data for PMOL, copolymers and hot-melt EF | | |
|---|---|---|
| Formulations | Glass transition/melting endotherm (°C) | Melting endotherm/enthalpy (°C /$\Delta$H, Jg$^{-1}$) |
| PMOL | – | 169.1/137.0 |
| SOL | 68.4 | – |
| VA64 | 105.5 | – |
| EF | | |
| PMOL/SOL (40%) | 70.8 | 141.3/9.7 |
| PMOL/SOL (50%) | 59.5 | 144.6/6.4 |
| PMOL/SOL (60%) | 81.8 | 150.8/11.6 |
| PMOL/VA64 (40%) | 52.9 | 137.0/13.3 |
| PMOL/VA64 (50%) | 105.8 | 131.3/7.2 |
| PMOL/VA64 (60%) | 102.2 | 123.3/5.5 |

The PMOL-VP/VA EXT showed two endothermic transitions: one close to the $T_g$ of bulk VP/VA (105 °C), and the other between 123–137 °C (melting transitions of PMOL Form II) depending on the PMOL loadings. The low temperature $T_g$ ranges from 52 to 102.22 °C based on the drug loadings. The calculated degree of crystallinity (data not shown) for all EF showed that the presence of crystalline PMOL is increased with the PMOL loading in each formulation. Similarly, the volume fraction of the PMOL is directly proportional to the amount of PMOL used and the melting transitions in the EF with SOL. Interestingly, a reverse effect was observed with the VP/VA polymer (**Figure 9.2**) where increased PMOL percentages resulted in lower drug melting points possibly due to the PMOL interactions in the VP/VA polymer matrix.

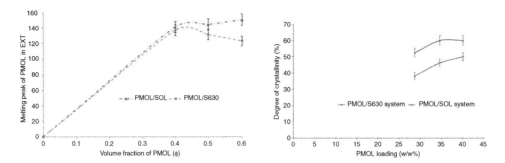

**Figure 9.2** Effect of PMOL volume fractions over the melting point

Fragility measures the rate which corresponds to the changes of various properties such as viscosity and thermal enthalpy relaxation/transition time, as a function of small difference in $T_g$ values. The fragility index (FI), also known as the stiffness index, *m* has been commonly used for characterising glassy materials. The FI range for strong to fragile (weak) glasses varies for $m < 100$ and $100 < m < 200$, respectively [9, 10]. The fragility of the crystalline PMOL in this study was calculated by extrapolating entropy values to zero. The activation enthalpy of the structural relaxation at the glass transition was estimated by using different scanning rates to achieve different $T_g$ and glass transition widths as expressed in **Equation 9.4**:

$$\ln q_+ = -\frac{E_a}{RT} \tag{9.4}$$

Where $E_a$ is the activation energy and $R$ is the gas constant and $q$ is the heating rate in DSC. As described in **Equation 9.4**, $E_a/R$ is the slope of the plot $1/T_g$ against $\ln (q)$ (*q* as heating rate in DSC) [28]. The $T_g$ is defined as the temperature of intersection between the equilibrium volume or entropy/temperature liquid curve and the linear extrapolation of the glassy curve [29].

$$m = E_a / (2.303 * R * T_{gm}) \tag{9.5}$$

By using **Equation 9.5**, the FI of PMOL (Form I) was found to be 83.8, which is lower than the amorphous PMOL estimated by Qi and co-workers 86.7 [10], suggesting that crystalline PMOL is a stronger glass compared to amorphous. Similarly, the *m*-values for PMOL in extruded PMOL-VP/VA formulations were found to be 97.7,

184.8 and 188.9 for 40–60% loadings, respectively. The calculated *m*-values of PMOL in EF are neither similar to the amorphous state nor to the crystalline Form I. This probably means that the extruded PMOL represents a more fragile system, which could be held together by weak isotropic bonds such as van der Waals interactions. It has also been reported that the strong glasses are typically composed of stronger, often covalent bonds that form three-dimensional networks [11, 30]. The foregoing indicates that the PMOL is not present in its original polymorphic form (strong glass) in the EF.

HSM studies were conducted to visually determine the thermal transitions and the extent of drug melting within the polymer matrices at different stages of heating. The bulk PMOL showed no change up to 100–170°C, which was distinctive in the DSC as no thermal transitions did occur until approximately 170.2 °C. Similar to the DSC thermograms, the PMOL/SOL EXT showed nominal API solubilisation until reaching temperatures of 140–142 °C and thereafter showed extensive melting of the drug. For the PMOL-VP/VA EXT it was noted that PMOL was extensively melted (and perhaps solubilised) in the polymer matrix at 128 °C. Assessment of the solubility parameters for the PMOL/SOL and PMOL-VP/VA composition also showed minimal difference, indicating similar interaction energies during mixing as discussed above, suggesting that the drug is well miscible with both polymers used.

The main aim of the overall study was to investigate whether VTXRPD can be effectively used as a predictive tool for the determination of polymorphic transformations of drug/polymer systems during HME processing. For this reason, the effect of temperature on the polymorphic transformations was investigated for bulk PMOL, as well as PMOL/SOL and PMOL/(VP/VA) PM. These findings were then compared with the VTXRPD profiles of the hot melt (HM) EF. The standard XRPD profile for Form I (monoclinic) PMOL showed distinct crystalline peaks at 2θ angles of 12.11, 13.82, 15.52, 18.20, 20.42, 23.51, 24.39 and 26.59° and a series of smaller (less intense) peaks at different 2θ angles ranging from 26.78–38.45° at ambient temperature. VTXRPD analysis clearly showed that PMOL polymorph Form I was stable from ambient temperature up to 160 °C (**Figure 9.3**), while no crystal transformation occurred within this temperature range. Further increase in temperature led to the transformation of PMOL to polymorph Form II (orthorhombic), which was complete at 165 °C (**Figure 9.3**). The characteristic crystalline peak at 24.36° 2θ for polymorphic Form I [31] started shifting above ambient temperature and was complete at 165 °C with a new peak position at 24.03° 2θ. This new peak displayed higher intensity than that at 24.36° and is strong evidence for the polymorphic transition from the monoclinic to the orthorhombic form.

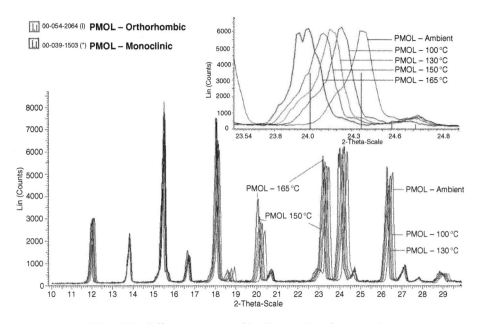

**Figure 9.3** VTXRPD diffractograms of bulk PMOL (from ambient to 165 °C)

Similar studies were undertaken for binary PM and EF of PMOL/SOL and PMOL/VP/VA, as shown in **Figure 9.4** (**Table 9.4**). It is evident that both the SOL and VP/VA play a key role in the PMOL polymorphic transformation where the critical transformation temperature is different to that of bulk PMOL. In the PMOL/SOL 40% PM, the characteristic peak at 24.4° 2θ starts shifting with an increase in temperature and completely transforms at 112 °C to a higher intensity peak at 24.0° 2θ, which corresponds to the orthorhombic form (**Figure 9.4**). A further increase in temperature up to 120 °C did not change the polymorphic form of PMOL (Form II) and then the XRPD peaks start reducing even with a small increase in temperature, as little as 1 °C. A similar thermal dependency was observed in the PMOL/SOL 40% EF (**Figure 9.4**). The presence of SOL resulted in a significant reduction of the PMOL transformation temperature to another polymorphic form. This observation is very important in relation to HME as it provides not only significant insights into the selection of extrusion temperature but also on the resultant state of the EXT.

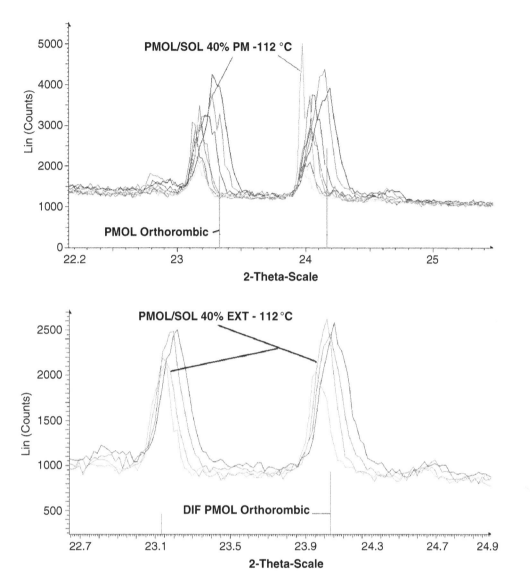

**Figure 9.4** VTXRPD diffractograms of PMOL/SOL 40% PM and EXT (from ambient to 122 °C). DIF: Diffraction index data

Further VTXRPD analysis was performed using PMOL/SOL 60% EXT formulations where the temperature was increased up to 122 °C and subsequently decreased to 100 °C in order to investigate possible polymorphic PMOL transformations (**Table 9.4**).

Interestingly, as a result of the temperature increase a complete transformation to Form I was detected at 120 °C (slightly higher than that of the PMOL/SOL 40% formulation) although the onset transformation temperature occurred at 112 °C. The reason for the slight increase in the transformation temperature could be attributed to the higher amount of PMOL present in the PMOL/SOL 60% formulation. When the temperature was decreased from 120 °C towards ambient, a complete reverse transformation occurred. This indicates that the polymorphic transformation of PMOL in PMOL/SOL systems is reversible as a function of temperature.

Studies were conducted with the EF in order to investigate whether similar changes take place as a function of temperature. In the case of PMOL-VP/VA 60% EF, the diffraction pattern (ambient temperature) suggested the existence of Form I but with the temperature increase, a clear transformation to the orthorhombic form was found at 120 °C and the characteristic diffraction peak shifted to 24.01° 2θ (**Table 9.4**). Similar observations have been found in other EF with different drug loadings.

| Table 9.4 Polymorphic transformation temperatures of formulations (PM and EF) | |
|---|---|
| **PM and EF (w/w %)** | **Temperature of polymorphic transformation (°C)** |
| PMOL/SOL (40%) | 112 |
| PMOL/SOL (50%) | Starts at 112, completes at 116 |
| PMOL/SOL (60%) | Starts at 112, completes at 120 |
| PMOL/VA64 (40%) | 120 |
| PMOL/VA64 (50%) | 120 |
| PMOL/VA64 (60%) | 120 |

The PM of the PMOL-VP/VA 40% formulation showed the transformation of PMOL at 120 °C which is higher than that of the PMOL/SOL 40% formulation. This higher temperature could be due to the relatively higher $T_g$ (~105 °C) of VP/VA polymer compared to the $T_g$ of SOL (~70 °C). The overall findings from VTXRPD analysis confirm that this technique could successfully be used as a predictive tool for extrusion processing and moreover to predict the possible temperature effects on the polymorphic transformation of crystalline drugs e.g., PMOL. A similar study was also conducted for the prediction and analysis of the polymorphic transformation of mebendazole Form C [32]. From the discussion above it is clear that the drug loading plays a vital role in the polymorphic transformation of PMOL Form I in PMOL/SOL combinations, while a discrepancy has been observed in PMOL-VP/VA formulations. Regardless of the drug loadings, polymorphic transformations occur at 120 °C in all PMOL-VP/VA formulations.

### 9.3.3 In-line Near-infrared Spectroscopy Monitoring

NIR spectra of PMOL Form I was measured off-line so as to determine the characteristic peaks attributable to the possible polymorphic transformation (**Figure 9.5**). Subsequently, in-line measured NIR spectra of the PMOL/SOL 60% formulation were collected from different mixing zones during HME. The data in **Figure 9.5** shows there is a significant intensity difference between the NIR spectrum of PMOL Form I and the PMOL in the EF. It has been reported in previous NIR studies [33] that the two polymorphs of PMOL exhibit different vibrational band profiles in the $5,700–6,400$ cm$^{-1}$ wavenumber region. Similarly, from the data in **Figure 9.5** it can be seen that the pre-processed 2$^{nd}$ derivative spectra showed a significant intensity difference, notably between $6,200$ and $6,400$ cm$^{-1}$, indicating that the polymorphic transformation of PMOL from its most stable Form I to the metastable Form II takes place during the HM extrusion process at 120 °C. From the results obtained it can be concluded that the intensity difference between the NIR spectra of PMOL Form I and the EXT (PMOL/SOL 40–60%) support the findings from the VTXRPD data with regard to the predictions of the polymorphic transformation of PMOL during HME.

## 9.4 Conclusions

In this study VTXRPD was successfully used as a predictive tool for possible polymorphic transformations of PMOL EXT from the water-soluble polymers SOL and VP/VA respectively. The monitoring of the polymorphic transformations of PM and EF at various temperatures through VTXRPD demonstrated that the stable Form I (monoclinic) transformed to Form II (orthorhombic) at temperatures above 120 °C. These findings were also supported by in-line NIR monitoring during the extrusion processing of PMOL/polymers formulations. In conclusion, VTXRPD can effectively be used as a predictive tool to monitor possible polymorphic transformation of active substances in solid dispersions and be a valuable technique for the development of EF by reducing processing times and efforts.

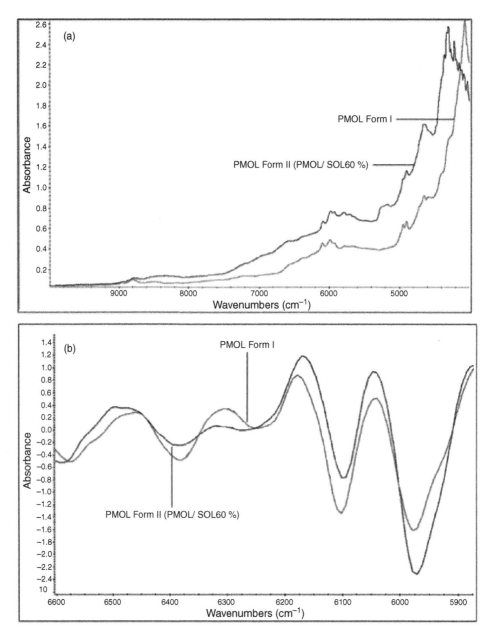

**Figure 9.5** (a) NIR spectra of PMOL pure Form I and PMOL EXT at 120 °C and (b) 2$^{nd}$ derivative NIR spectra of SOL, PMOL Form I and PMOL EXT

## References

1.   K. Sekiguchi and N. Obi, *Chemical and Pharmaceutical Bulletin*, 1961, **9**, 866.

2.   P. Srinarong, H.D. Waard, H.W. Frijlink and W.L. Hinrichs, *Expert Opinion on Drug Delivery*, 2011, **8**, 1121.

3.   M.A. Alam, R. Ali, F.I. Al-Jenoobi and A.M. Al-Mohizea, *Expert Opinion on Drug Delivery*, 2012, **9**, 1419.

4.   D.J. Jang, T. Sim and E. Oh, *Drug Development and Industrial Pharmacy*, 2013, **39**, 7, 1133.

5.   I. Kushida and M. Gotoda, *Drug Development and Industrial Pharmacy*, 2012, 38, 10, 1200.

6.   X. He, L. Pei, H.H. Tong and Y. Zheng, *AAPS PharmSciTech*, 2011, **12**, 104.

7.   K. Gong, R. Viboonkiat, I.U. Rehman, G. Buckton and J.A. Darr, *Journal of Pharmaceutical Sciences*, 2005, **94**, 2583.

8.   M. Maniruzzaman, M.M. Rana, J.S. Boateng, J.C. Mitchell and D. Douroumis, *Drug Development and Industrial Pharmacy*, 2013, **39**, 218.

9.   M. Maniruzzaman, J.S. Boateng, M. Bonnefille, A. Aranyos, J.C. Mitchell and D. Douroumis, *European Journal of Pharmaceutics and Biopharmaceutics*, 2012, **80**, 433.

10.  S. Qi, P. Avalle, R. Saklatvala and D.Q.M. Craig, *European Journal of Pharmaceutics and Biopharmaceutics*, 2008, **69**, 364.

11.  S. Qi, A. Gryczke, P. Belton and D.Q.M. Craig, *International Journal of Pharmaceutics*, 2008, **354**, 158.

12.  S.K. Rastogi, M. Zakrzewski and R. Suryanarayanan, *Pharmaceutical Research*, 2001, **18**, 267.

13.  J.M. Rollinger and A. Burger, *Journal of Thermal and Analytical Calorimetry*, 2002, **68**, 361.

14.  S.K. Rastogi, M. Zakrzewski and R. Suryanarayanan, *Pharmaceutical Research*, 2002, **19**, 1265.

15. Y. Li, J. Han, G.G. Yang, D.J. Grant and R. Suryanarayanan, *Pharmaceutical Development and Technology*, 2000, 5, 257.

16. C.M. Hansen, *Industrial Engineering and Chemistry Research Development*, 1969, 8, 2.

17. P.J. Hoftyzer and D.W.V. Krevelen in *Properties of Polymers*, Elsevier, Amsterdam, The Netherlands, 1976.

18. T. Nishi and T.T. Wang, *Macromolecules*, 1975, 8, 909.

19. J. Hildebrand and R. Scott in *Solubility of Non-electrolytes*, 3rd Edition, Reinhold, New York, NY, USA, 1950.

20. PDF-2 Release in Kabekkodu SN, International Centre for Diffraction Data, Newtown Square, PA, USA, 2008.

21. F.J. Hempenstall, I. Tucker and T. Rades, *International Journal of Pharmaceutics*, 2001, 226, 147.

22. X. Zheng, R. Yang, X. Tang and L. Zheng, *Drug Development and Industrial Pharmacy*, 2007, 33, 791.

23. J.S. Higgins, J.E.S. lipson and R.P. White, *Philosophical Transactions of the Royal Society A*, 2010, 368, 1009.

24. A. Paudel, E. Nies and G. Van den Mooter, *Molecular Pharmaceutics*, 2012, 9, 3301.

25. Y. Zhao, P. Inbar, HP. Chokshi, W. Malick and D.S. Choi, *Journal of Pharmaceutical Sciences*, 2011, 100, 3196.

26. S. Janssens, A.D. Zeure, A. Paudel, J. van Humbeeck, P. Rombaut and G. van den Mooter, *Pharmaceutical Research*, 2010, 27, 775.

27. M. Maniruzzaman, D.J. Morgan, A.P. Mendham, J. Pang, M.J. Snowden and D. Douroumis, *International Journal of Pharmaceutics*, 2013, 443, 199.

28. J.M. Barton, *Polymer*, 1969, 10, 151.

29. B. Hancock, C. Dalton, M. Pikal and S. Shamblin, *Pharmaceutical Research*, 1998, 15, 762.

30. A. Rossi, A. Savioli, M. Bini, D. Capsoni, V. Massarotti, R. Bettini, A. Gazzaniga, M.E. Sangalli and F. Giordano, *Thermochimica Acta*, 2001, 406, 55.

31.  P.D. Martino, P. Conflant, M. Drache, J.P. Huvenne and A.M. Guyot-Hermann, *Journal of Thermal and Analytical Calorimetry*, 1997, **48**, 447.

32.  M.M. De Villiers, R.J. Terblanche, W. Liebenberg, E. Swanepoel, T.G. Dekker and S. Mingna, *Journal of Pharmaceutical and Biomedical Analysis*, 2005, **38**, 435.

33.  I.C. Wang, M.J. Lee, D.Y. Seo, H.E. Lee, Y. Choi, W.S. Kim, C.S. Kim, M.Y. Jeong and G.J. Choi, *AAPS PharmSciTech*, 2011, **12**, 764.

# 10 From Pharma Adapted Extrusion Technology to Brand New Pharma Fitted Extrusion Design: The Concept of Micro-scale Vertical Extrusion and its' Impact in Terms of Scale-up Potential

Victoire de Margerie and Hans Maier

## 10 Introduction

Hot-melt extrusion (HME) technology is well-known by the food industry (pasta, chocolate) and also widely used for producing/shaping many materials (plastics, aluminium, composites), but it has only been of interest for the pharmaceutical industry since the early 2000s [1–10]. HME is a multi-step process where the main material gets into the machine through a feeder and is then combined with various ingredients/additives in order to create the final product. Raw materials progress into a metal tube (the barrel) where they are very precisely mixed and heated into a liquid state (the melt), thanks to the screws included into the barrel that create the 'mashing' effect. The melt is then pushed through a die (that gives the final shape) and cooled down (through air or water cooling), making the product available for further downstream processing.

HME can therefore be used for embedding drugs into polymeric carriers with the help of processing aids (plasticisers and antioxidants). We think that the technology offers various advantages over the traditional pharmaceutical processes [2, 3, 11–16]. Adapting HME to the pharmaceutical environment is however not straightforward. Upscaling results achieved at lab-scale also requires in depth work and know-how. As a result, HME is still not a technology broadly used by the pharmaceutical industry. Our conviction is that a new technology breakthrough is needed and this is vertical extrusion.

## 10.1 The Advantages of Hot-Melt Extrusion

HME has been tested/used in various shapes (pellets, granules, tablets, capsules, sheets, powder and so on) and for various purposes e.g., masking the bitter taste of a drug, formulating targeted-release/controlled-release dosage forms [3]. There are numerous advantages of HME techniques in pharmaceutical product manufacturing and development (summarised in **Table 10.1**).

| Table 10.1 Features and advantages of hot-melt extrusion in continuous process | |
|---|---|
| Feature | Benefit |
| Continuous process | Economical, efficient scale-up |
| Multiple batch operations | Precise dosing, melting, mixing, degassing, and shaping materials in one single process in a short-time |
| Solvent free processing | Economical, no residual solvent in final product, thus waiving need for control stages |
| Intense mixing and agitation | Improved content uniformity |
| Process analytical technology | FDA's process analytical technology initiative readily applied, less off-line testing |
| Extensive automation | Increase productivity, consistent quality products according to good automated manufacturing practice |
| Processing of thermally sensitive API | Use of ports downstream on the extruder barrel |
| Solubility enhancement | formation of solid dispersions |
| FDA: US Food and Drug Administration | |

### 10.1.1 Examples of Problems Solved Thanks to Hot-Melt Extrusion

HME provides solutions for the following issues (all figures are given in GBP):

• Active pharmaceutical ingredient(s) (API) sometimes are very expensive (above £1,000 a gram) and therefore benefit from being processed on micro-scale machines in order to minimise and control lot size (50 g minimum lot size for a 10 mm extruder *versus* 1.350 g for a 30 mm extruder). Hence, trial costs involving those API (if one 2 kg trial on a dosage form that includes 10% of a £5,000 per g API costs £1 million, who would want to conduct a series of hundreds of trials to bring a new formulation for FDA approval and further industrialisation? It becomes much more feasible with a trial of 100 g offering the same precision at a £5,000 cost and so on).

- Both pharmaceutical grade polymers and API are difficult to process, being usually sticky, temperature sensitive and shear sensitive materials. They, therefore require a consideration of all process parameters (screw configuration, screw speed, melt pressure and so on) or ingredients (plasticisers, $CO_2$ and so on) in order to lower the shear forces needed or the temperatures required to properly mix the final product, while avoiding thermal degradation.

- On top of lowering those process parameters, controlling them with the smallest possible deviation is crucial to optimise the interference of the API with the functionality of the other components in the formulation (e.g., Vitamin E TPGS has been reported to plasticise polyethylene oxide and enhance drug absorption by suppressing its melting point). Therefore, it requires precise dosing of all product ingredients and metering of all process parameters.

- Pharmaceutical class extruders must also meet regulatory requirements – the metallurgy of the contact parts must not be reactive, additive or absorptive with the product and the equipment must be configured for the cleaning and validation requirements associated with a pharmaceutical environment [so called good manufacturing practices (GMP)].

## 10.2 From Micro-scale to Industrial-scale

Since 2007, Rondol has been a pioneer in developing micro-scale equipment for pharmaceutical research and installed the first twin-screw extrusion 21 mm industrial line for pharmaceutical production in 2014. The installation of this industrial line followed a 2-year experimental process in order to scale-up results achieved at lab-scale with a 10 mm twin-screw extruder.

Using differential scanning calorimetry, three samples have been characterised by heating, cooling and reheating: one with the polymer alone, one with the polymer compounded with 5% API and one with the polymer compounded with 11% API. For the polymer alone, at the first heating phase we noted a hysteresis at glass transition temperature ($T_g$) = 45.46 °C but no other one. The same results are obtained at the reheating phase. For the polymer with 5% additive, at the first heating phase, we noted a hysteresis at $T_g$ = 44.54 °C and another hysteresis between 156 and 165 °C. At the reheating phase, we can see a notable difference on the hysteresis at 165 °C that is now disappearing. It tends to prove that the API is destroyed at temperatures over 156 °C. For the polymer with 11% additive, at the first heating phase, we noted a hysteresis at $T_g$ = 49.67 °C and another hysteresis between 156 and 166 °C. At the reheating phase, we can see a notable difference on the hysteresis at 156 °C that is now disappearing. It tends to prove that the API is destroyed at temperatures over 156 °C.

*1st conclusion*: We need to favour mechanical action (mixing) and avoid putting too much heat into the compound as it would destroy the API. Therefore, we would choose a twin-screw extruder that can provide efficient mixing, rather than a single-screw extruder that is merely a pressure generating device.

### 10.2.1 Viscosity of the Three Samples at Different Temperatures

For all three samples, the importance of a reduced residence time within the extruder was made clear. When residence time increases, polymer chains start to break which is evident by the lower measurement on viscosity. This means that the polymer is degrading when residence time increases. The light grey line of **Figure 10.1** clearly shows that the viscosity decreased because material stayed in the barrel (otherwise, the light grey line would be superimposed to the dark grey line).

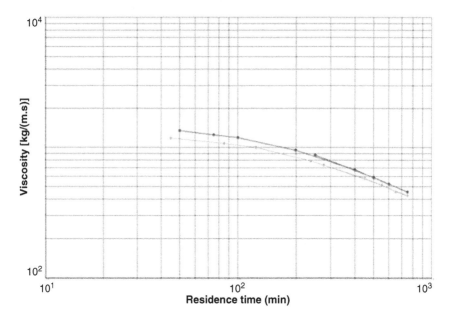

**Figure 10.1** Residence time increase *versus* the measurement on viscosity

*2nd conclusion*: We would recommend an optimum length/diameter (L/D) ratio (barrel length) that offers both high quality of mixing/dispersion and short residence time; 30:1 for a 21 mm twin-screw extruder.

### 10.2.2 Viscosity at Various Temperatures

The percentage of API used in the compound also impacts the viscosity of the materials. The survey showed that sample B (11% additive) is a lot more viscous than sample A (5% additive), which is itself more viscous than the polymer alone:

- Polymer alone: 325.4 Pas

- Polymer with 5% additive: 583 Pas

- Polymer with 11% additive: 3,027 Pas

So we can see that the additive increases the viscosity a lot. This increase is not linear, therefore the product will become more and more difficult to process when API loading increases. The polymer with 11% additive will be a lot more difficult to process. You would need either to increase the temperature (but it would quickly destroy the API) or to increase the mechanical work applied to the compound (thanks to the efficient mixing provided by a twin-screw extruder). The flexible screw configuration will further enhance this efficient mixing without increasing either pressure or temperature. Also at a given temperature, pressure and screw configuration, a larger twin-screw extruder will provide better mixing because it has more free volume and it can handle small pellets better.

*3rd conclusion*: A 21 mm twin-screw extruder is recommended rather than a 10 mm twin-screw extruder when you need to process various compounds with varying API percentages.

Our PASS software has been used to simulate:

- The optimum screw configuration for all three compounds.

- The optimum screw speed and temperature profile for each of those configurations (it has to be fast enough to minimise the residence time but not faster than necessary to avoid increased shear forces).

Two examples are given below as an illustration.

We tested one option of light mixing for the sample with 5% additive (with a combination of 30° and 90° mixing blocks) and obtained the following result: with a screw speed set-up at 100 rpm, we have a mean residence time of 48 s and we never exceed a melt temperature of 125 °C, with a minimum at 115 °C (so we stay far away from the 156 °C degradation point), see **Figure 10.2.**

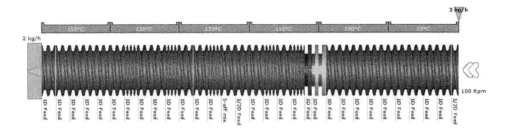

Figure 10.2 Screw configuration and set-up 1

We also tested what we consider to be the optimum ultra-light mixing for the sample with 11% additive (with 30° mixing blocks) and obtained the following result: with a screw speed set-up at 100 rpm, we have a mean residence time of 47 s and we never exceed a melt temperature of 133 °C, with a minimum at 119 °C (so we stay far away from the 156 °C degradation point), see **Figure 10.3**.

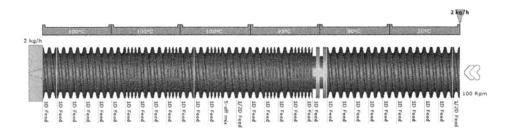

Figure 10.3 Screw configurations and set-up 2

## 10.3 Hot-Melt Extrusion as a Standard Technology for the Pharmaceutical Industry

Over the last few years HME has been adopted as a standard technology by the pharmaceutical industry. Development and technical revolutions using in-house expertise and facilities at Rondol, France, have so far produced various HME machines at different scales available in commercial spheres:

- Already in 2007, we sold to a major pharmaceutical company the first commercial 10 mm twin-screw extruder (minimum lot size between 50 and 100 g). Since then, we have sold more than 50 machines worldwide – academic endorsements include Queen's University in Belfast, the Institute für Nanomaterialien in Saarbrücken, the University of Newcastle or the University of Istanbul.

- In 2008, we offered a major pharmaceutical company an 'easy clean' barrel that allows quick access to screws and barrel for cleaning, screw configuration change and unrestricted wipe-down thanks to a removable barrel liner, therefore leaving a minimal residual quantity in the barrel at the end of a production run (less than 10 g).

- In 2010, we set-up for a major pharmaceutical company this same barrel with a length to screw diameter ratio of 40 (the standard ratio is 20) which allows the 'soft' extrusion described above in order to avoid the degradation of API.

- In 2011, we offered a major pharmaceutical company an optional high temperature operation of up to 450 °C that enables the machine to process polymers such as polyether ether ketone or polyether ketone ketone.

- In 2012, we opened our R&D facility in Strasbourg, highlighting the development of new machines for pharma and medtech applications.

- In 2013, we successfully completed a study highlighting the formation of solid foams for oral applications by HME with the objective of increasing the release rate of the final product. $CO_2$ was used both as a cooling and as a blowing agent, directly injected into the barrel during extrusion.

- In 2014, we scaled-up on to a 21 mm GMP extrusion line the results achieved on our 10 mm lab extruder and have installed the first industrial twin-screw extrusion line.

But, we need to recognise that the specific advantages offered by HME (more efficient for the patients, more easily compliant with FDA requirements and less capital intensive for the industry) are not well-known by top deciding bodies (CEOs of pharma groups, healthcare organisations or public regulatory entities). Work still needs to be done to further minimise the quantities of API required, further improve the versatility of the extrusion process, further extend the range of ancillaries and downstream equipment (e.g., Rondol has also developed a microencapsulation unit), and make the GMP compliance more user friendly.

## 10.4 Innovations

The next innovation challenge is to speed up a broader use of HME in the pharmaceutical industry while still keeping it 'low cost', therefore allowing for efficient homecare at affordable costs. We think that vertical micro-scale extrusion could well be the breakthrough that will increase the extrusion market size and enable HME to become a standard pharmaceutical technology.

The current horizontal format is indeed a bottleneck to various improvements:

- *Contamination issues*: The equipment footprint of horizontal extrusion machines is quite large and it is subjected to high risks of contaminations from the work environment to the API and from the API to the work environment. This is especially critical in the manufacture of highly potent drug materials, e.g., therapeutica for the treatment of cancer.

- *Cleaning issues*: A horizontal extruder is quite big, especially with a 40/1 barrel, which makes the cleaning both long and fastidious. Cleaning is central to pharmaceutical processing as both the equipment and the manufacturing room must be fully cleaned after every production run. This therefore has a huge impact on efficiency and costs.

- *Operator interface*: A big footprint also means a more complex operator interface due to numerous moves along the extrusion line, e.g., along the in-feed of the filler *via* the control panel to the material ejection at the pelletiser.

Based on a market analysis and a vision for the future of pharmaceutical HME, the preferred process flow direction needs to be changed from the horizontal position to top-down in the vertical axis. As a consequence, we are prototyping a fully vertical micro-extruder line with co-rotating and parallel screws and a fully integrated automation system, (both characteristics are essential to successful scale-up for industrialisation purposes).

The following scheme illustrates the vertical extruder in its prototype format (**Figure 10.4**):

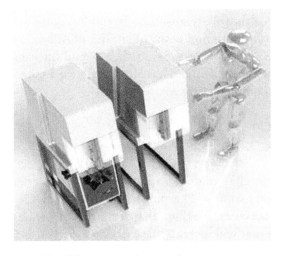

**Figure 10.4** The vertical extruder in its prototype

Designed for pharmaceutical applications, the vertical extruder will take into account the well-known technical and economic constraints linked to pharmaceutical applications (the capability to produce small lots and the necessary 'soft' extrusion process).

The vertical format will reduce the equipment footprint and the vertical in-feed of the strand into any of the ancillary downstream equipment, e.g., pelletisers or calenders to eliminate the risk of contamination from the use of horizontal working conveyors Cleaning will also be simplified as Rondol  improves the current 'easy clean' system.

The table top design of the vertical extrusion system also allows modular integration of all ancillary equipment necessary for final product processing and shaping (pelletisers, calenders, milling equipment and so on) below the extruder. In the traditional horizontal extrusion technology, an array of ancillary equipment is used for further downstream processing, comprising a conveyor belt combined with a pelletiser. This typically results in a long process chain.

Operator working conditions will be improved as there will be no need to move around the equipment in order to control process performance.

Last but not least, the vertical position is largely able to cover and seal the extruder to protect operators and the working environment from hazardous vapours/dusts and the pharmaceutical ingredients/excipients from contaminations from the outer environment (**Figure 10.5**).

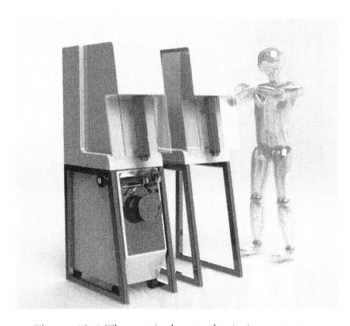

**Figure 10.5** The vertical extruder in its prototype

In **Table 10.2**, the advantages and disadvantages of horizontal and vertical processing are compared and contrasted.

| Table 10.2 Comparison of aspects for horizontal and vertical extrusion processing | | |
|---|---|---|
| **Aspect** | **Vertical** | **Horizontal** |
| Footprint | Small | Medium |
| Contamination risk | Very low | Medium |
| Cleaning | Easy | Medium |
| Integration of ancillary equipment | Vertical plug-in of modular downstream process equipment | Horizontal array of downstream equipment |
| Process control through operator | Easy | Medium |
| Costs | Fair | Average |

The advantage of vertical processing is not limited to, but is preferred for small-scale table top extruders with process routes ranging from 200 to 850 mm (e.g., 10 mm twin-screw extruders from 20 L/D to 21 mm extruders to 40 L/D); these extruders are typically those requested by the pharmaceutical industry for processing complex and costly API both at laboratory and industrial scales. All necessary drive and control elements, such as the drive and the cooling system, as well as the electrical supply and the data system is integrated into the table top extruder casing (**Table 10.2**).

In principle, the operational system of the vertical extruder is based on the same design geometry as its horizontal homologues so that the positive conveying mechanism of the fully intermeshing co-rotating twin-screws provides predictable scale-up from vertical small-scale machines to horizontal large-scale extruders, as long as both machines are of the same geometry, e.g., co-rotating and fully intermeshing.

The following technical issues are being addressed:

• The necessity to design a generic extruder capable of integrating by an easy 'plug in' mechanism, all types of top and side feeders for the API and further solids, liquids and gases e.g., $CO_2$.

• The work on all screw elements design and configuration in order to provide smooth additive transportation through the secondary feeders that are now horizontal and so on.

- The simplification of the extruder 'easy clean' system (see above).

- The integration of a high performance cooling systems for strand cooling.

- The integration of downstream equipment with versatile technical constraints (temperature control for calender or cutting precision for pelletiser).

- The coordination of all process parameters thanks to an integrated, user friendly and reliable automation system.

## 10.5 Conclusions

Even in horizontal format, HME has already demonstrated its capability to offer specific advantages in order to address challenging pharmaceutical applications (more efficient for the patients, more easily compliant with FDA requirements and less capital intensive for the industry). Vertical parallel twin-screw extrusion now seems to be the new frontier and the breakthrough that will expand the pharma extrusion market size and allow HME to become a standard pharmaceutical technology.

## References

1.  J. Breitenbach, *European Journal of Pharmaceutics and Biopharmaceutics*, 2002, 54, 107.

2.  M. Maniruzzaman, J.S. Boateng, M.J. Snowden and D. Douroumis, *International Scholarly Research Notices: Pharmaceutics*, 2012, Article ID: 436763.

3.  M.M. Crowley, Z. Feng, M. Repka, S. Thumma, S. Upadhye, S. Kumar, J. McGinity and C. Martin, *Drug Development and Industrial Pharmacy*, 2007, **33**, 909.

4.  M. Charlie, *Pharmaceutical Technology*, 2008, **32**, 10, 76

5.  G.P. Andrews, S. David, A.M. Osama, N.M. Daniel and S. Mark, *Pharmaceutical Technology Europe*, 2009, **21**, 1, 24.

6.  M. Crowley, B. Schroeder, A. Fredersdorf, S. Obara, M. Talarico and S. Kucera *International Journal of Pharmaceutics*, 2004, **271**, 1–2, 77.

7.  A. Forster, J. Hempenstall, I. Tucker and T. Rades, *Drug Development and Industrial Pharmacy*, 2001, **27**, 6, 549.

8.  M. Fukuda, N. Peppas and J. McGinity, *Journal of Controlled Release*, 2006, **115**, 2, 121.

9.  U. Hanenberg, O. Schinzinger, H. Maier and S. Junkering, *Pharmazeutische Industrie*, 2013, **75**, 2, 319.

10. H. Maier, *Drug Development & Delivery*, 2012, **12**, 8, 55.

11. D.A. Miller, W. Yang, R.O. Williams and J. McGinity, *Journal of Pharmaceutical Sciences*, 2006, **96**, 361.

12. B. Murray and R. Hilder in *Proceedings of the 4M International Conferences on Micro Manufacturing*, Forschungszentrum Karlsruhe, Germany, 2009.

13. C. Young, M. Crowley, C. Dietzsch and J. McGinity, *Journal of Microencapsulation*, 2007, **24**, 1, 57.

14. Y. Zhu, A. Malick, M.H. Infeld and J. McGinity, *Drug Development and Industrial Pharmacy*, 2006, **32**, 5, 569.

15. Y. Zhu, A. Malick, M.H. Infeld and J. McGinity, *Journal of Drug Delivery Science and Technology*, 2004, **14**, 4, 313.

# 11 Continuous Manufacturing *via* Hot-Melt Extrusion and Scale-up: Regulatory Aspects

Mohammed Maniruzzaman

## 11 Introduction

As an emerging processing technology, hot-melt extrusion (HME) has recently been successfully exploited for the formulation development and optimisation of various poorly water-soluble or insoluble drugs. By using an HME based processing technique, generally a stable amorphous solid dispersion is achieved that results in an improved dissolution rate and thus increased oral bioavailability. Therefore, HME has been successfully applied to develop multiple drug delivery systems and to the manufacture of pharmaceutical drug products over the last three decades or more [1–4]. The numbers of drugs being approved by the US Food and Drug Administrations (FDA) using HME include drugs across the whole range of the therapeutic classes and include a range of dosage forms such as tablets, strips/films, capsules and so on. Moreover, solvent free extrusion processing can be managed using environmentally viable equipment complying with processing/manufacturing standards required for the pharmaceutical manufacture at a potentially high production rate. HME also reduces the energy, plant and carbon footprint as this emerging technique can be adopted as a continuous manufacturing platform. From a background viewpoint, HME based applications applied to pharmaceutical manufacturing dates back to the 1930s when J. Brama mainly started the use of HME in the plastics [5] and food industry [6].

## 11.1 Continuous Manufacturing and Hot-Melt Extrusion Processing Technology

In recent years, there has been a rising interest by the pharmaceutical industry for pharmaceutical product manufacture, design and optimisation involving continuous manufacturing processes [7]. In reality, the conventional batch processes of manufacturing are not particularly good for product quality assurance which renders a number of drawbacks such as poor controllability, low yield, and difficult scalability. In addition, batch processes are very labour intensive and typically exhibit low plant productivity [8]. Thus, ongoing efforts have focused on the development and optimisation in continuous processes. However, in the pharmaceutical industry, the application of continuous manufacturing, automation and control is still a challenging task. Therefore, there is an immense need to overcome some challenges involved in continuous manufacturing prior to the design and implementation of the process.

Generally, in a continuous manufacturing process, feeding of the input materials or mixtures (known as starting materials/raw materials such as polymers, drugs, lipids or inorganic excipients) and the evacuation of processed output materials take place concurrently in a continuous mode. In a fully automated continuous manufacturing process controlled/automated by software, different steps involved in the processing of the materials are coordinated to form a continuous production flow whilst the finished product collection occurs simultaneously (preferably at the same rate as the throughput of the raw materials in the feeding sections).

As previously mentioned, in a traditional batch approach, pharmaceutical manufacturing is really a series of disconnected single unit operations, usually installed in separate rooms, with product waiting to be released to the next process step in another room. Thus, there is a tremendous need to make the equipment completely interconnected hardware-wise with a user friendly software interface to operate the hardware as one line. It can then be expected that the whole line is operated by a single operator. A continuous manufacturing platform will build up on a concept that can integrate with the equipment that precedes it and the one that comes next in the line. Such an ideal solution for the automation of the continuous manufacturing set-up can be, for an example, commercial Siemen's PAT (SIPAT) and supervisory control and data acquisition (SCADA) systems providing an integrated and data-driven control platform.

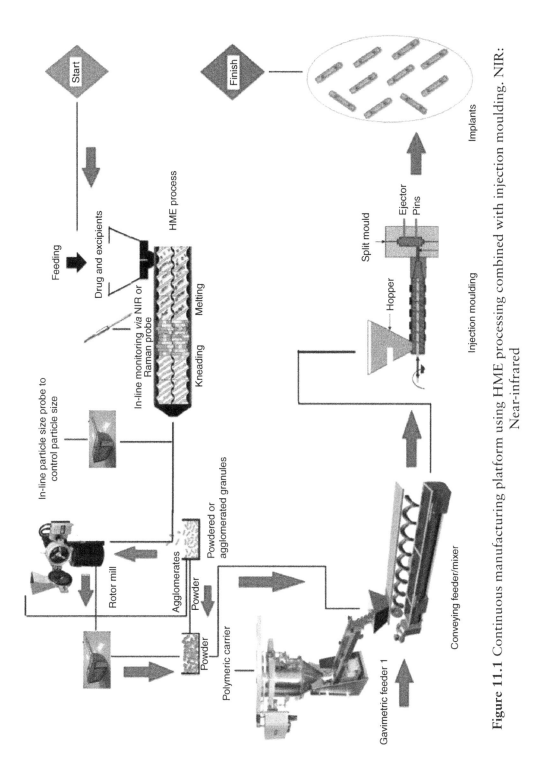

**Figure 11.1** Continuous manufacturing platform using HME processing combined with injection moulding. NIR: Near-infrared

Although it can be claimed that the HME process by itself is a continuous process and can be adopted for continuous pharmaceutical applications with an ease to up-scale for pilot scale manufacturing (**Figure 11.1**), in current pharmaceutical manufacturing, the existence of this emerging HME technique in a continuous automated format is not yet fully optimised. In a traditional HME based manufacturing process and operation, the extrudates are typically collected off-line with a set of preset processing parameters (e.g., temperatures, screw speed and feed-rate). The collected extruded materials (either in strand forms or agglomerated forms/powdered forms) are further subjected to a variety of downstream processing (e.g., pelletising, milling and blending with other excipients), followed by a series of off-line physicochemical characterisations (e.g., differential scanning calorimetry, X-ray diffraction, dissolution).

## 11.2 Aspects of the Controls/Parameters in Hot-Melt Extrusion Processing

As starting materials for HME processing continuously flow and are conveyed through the extruder barrel, frequent variations and deviations throughout the process are more likely to be observed. In those observed processing conditions, the materials used and/or the environmental factors may have a potential impact on the final product quality and performance over a determined period of time. Therefore, it is always a matter of significant importance to consider continuous operation protocols and parameters when developing a control portfolio for a continuous manufacturing process and operation. In an optimised continuous manufacturing platform, HME variation considerations may include both short term factors such as the feeding defects, screws disturbances as well as relatively longer term factors such as barrel/die wear. The existence of both the short-term and long-term pitfalls in the HME process necessitate the need for a robust, steady and consistent control policy to ensure product quality assurance and consistent quality attributes of the product processed over the total time span of the whole manufacturing platform. There is a set of crucial elements that can be addressed and included as concepts of development and implementation in the control policy for a fully continuous manufacturing technique *via* HME. These aspects can include for example, the state and nature of the control itself and/or the critical parameter, uniformity of the finished products and so on during an extrusion process.

In order to manufacture finished pharmaceutical products of uniform quality, performance and characteristics, a suitable set of experimental protocol has to be in place and predefined at the beginning of the experimental works that can isolate those products that do not comply with the set standard and quality. This can only be possible to identify and isolate, if all required analysis of product quality and variation of process parameters over time have been appropriately undertaken. Obviously, the sampling frequency will combine with the application of statistical methods to determine uniformity of quality measurements in order to differentiate the fully continuous manufacturing process over the frozen batch mode processing in terms of the product quality attributes. The establishment of appropriate limits of the quality in terms of the acceptance limit is a challenging task and is typically based on product performance understanding (e.g., qualification of impurities).

In HME processing, the ability to optimise the traceability of the materials during the process requires a thorough understanding of the materials' flow along the continuous process line under one operation control. These starting materials are fed in an HME process over a predefined time. Therefore, material traceability involves understanding the state of the process parameters during the time the final products are manufactured as a little deviation at any point in the process, may have a severe impact on the downstream quality and processes. It is very important that a thorough understanding of the flow of materials should be established prior to the system being set up. Experiments for the determination of the residence time distribution (RTD) for implementing the traceability of the materials conveyed through the extruder barrel [9] and parameters that have an influence on the process parameters on the RTD of the material within the extruder has been described elsewhere [9]. Similarly, the specific mechanical energy consumption and the possibilities of up scaling this process from a lab-scale to a production line extruder has also been described in the literature [9].

## 11.3 Continuous Manufacturing *via* Hot-Melt Extrusion

At present, the current scenario relating to pharmaceutical processing in the manufacturing environment is mainly referred to as a frozen batch processing method where each unit dosage form is identified by a unique single batch within which significant batch to batch variation can often be seen. This batch manufacturing approach has been used for decades and has been well recognised by both the industry and the regulatory bodies. Whereas various other manufacturing industries such as the petrochemical, chemical, polymer and food industries have steadily showed a trend to move towards the continuous manufacturing process, the pharmaceutical industry is still yet to realise the need to upgrade its production technology and adopt a continuous manufacturing process. This will have to be driven by both cost and quality considerations. A study [10] comparing batch *versus* continuous processing broadly discussed reasons why the pharmaceutical industry has traditionally been dominated by batch processing to date. The lack of flexibility, intensive costs and time taken for the developments are the major reasons of many in batch processing as to why other industries have moved to continuous processing processes. Other important factors that influence the decision to move from batch to continuous processing include, but are not limited to, the desire to minimise the required size of new manufacturing plants and the need to efficiently use the available capacity in-house [10]. However, both the pharmaceutical industry and the FDA have recently agreed that a re-evaluation of the manufacturing regulations applying to continuous manufacturing will prove itself to be beneficial for the end-user patient that both of them serve [11]. A spotlight discussion published in a recent article in the *Wall Street Journal* [12] served to highlight and illustrate the effects of these changes, the report shows the industry and the FDA are working together in various initiatives to implement new methodologies for quality testing and analysis.

Since the beginning of the new scientific era, technically, pharmaceutical companies have always competed with a special focus on innovation through new drugs for medical needs. A recent review of drug development costs outlined that approximately a huge amount of US $802 million is estimated to be spent on bringing a new drug to market. [13]. In addition, ongoing R&D costs will further increase the cost of drug development. Owing to various factors such as increases in competition, further increases in the proportion of generic drug product utilisation, the opening of new markets, and the socioeconomic pressures for price controls, it is envisaged that the industry has to look at some point for another alternative to reduce costs effectively. The overall shifts and future trends in the area of pharmaceutical innovative manufacturing and technologies will translate into manufacturing many more new products. The same cost and quality drivers which have been instrumental in other industries taking up a continuous manufacturing process are also forcing the pharmaceutical industry now to look for ways to improve quality whilst not compromising the product quality and performance but providing reduced manufacturing costs [12]. Continuous processing technologies will provide a possible path forward for the industry to reduce the cost of manufacturing by converting processes and steps from batch to continuous mode along with appropriate real time monitoring and analysis using state-of-the-art process analytical technologies (PAT) tools for the implementation of quality-by-design (QbD) – a new terminology being promoted by the FDA.

With consideration of appropriate real time monitoring and analysis using state-of-the-art PAT during a continuous manufacturing process such as HME, there are several approaches in the literature that have been reported in order to characterise RTD (which is considered an important factor for continuous manufacturing and scale-up) in the processing steps involved in extrusion processing in a manufacturing sphere [14–16]. One of the aspects of process development in a continuous mode of operation is the thorough analysis and understanding of RTD *via* the processing steps and the descriptions of the degree of intermixing along the processing steps. Another key point to consider when designing and optimising a continuous manufacturing process as well as the scale-up of HME, is the implementation of the overall manufacturing throughput or feed-rate. The change in the rate of the throughput/feed can have fundamental changes in various events inside the extruder barrel such as shear, mixing, and heating which normally impact on the material quality attributes and the blending for final product processing. So it can be said that the change in the throughput in a continuous operation of the HME process plays a vital role and forms an important part in the development of the overall control policy enforced by the regulatory bodies.

## 11.4 Continuous Manufacturing Process over Batch Process

The general implemented process used in the current manufacture of drug products consists of a series of unit operations. In this series, each operation is intended to control certain properties of the material being processed. Several of these commercially used unit operations such as tableting is commercially used in unattended operation and is a continuous compaction operation which is normally run in batch mode. Technically, any equipment that operates on a first in/first out principle can be considered continuous by design and therefore has the issue of start-up and shutdown, but operates at a steady state for most of the processing time. Materials processed *via* such equipment may have the same level of energy input, regardless of batch size.

One of the major advantages of the implementation of a continuous processing technology is that the scale and physical size of the batch do not need anywhere near the changes in equipment required by a batch processing system where the equipment frequently changes with increases in the scale and physical size of the production. In this way, batch manufacture involves the changing of the scale of the equipment as the batch size increases so that sometimes dramatic changes in equipment surface area to volume can occur during scale-up. This often leads to significant differences in what the product experiences in the manufacturing.

Based on the currently adopted technology viewpoint, the manufacture of solid dosage forms is carried out by using mainly three major methodologies. The first and simplest of them is direct compression which involves the blending with excipients followed by tableting. A continuous direct compression system may comprise several individual powder feeders to introduce the materials into a continuous blender/mixer where the last section of this process will feed the blended powder to a tablet press as a last stage of the tablet manufacturing. Next, slightly more involved, is the dry granulation process where the active and selected excipients are processed *via* roller compactor equipment, followed by a mill. The milled material is blended with suitable excipients and consequently tableted. The most involved, and the most common situation includes wet granulation with various unit operations involved in order to produce a solid oral dosage form. To successfully manufacture a solid oral dosage form using a continuous process, it is necessary to conduct wet granulation, drying, milling, blending and tablet coating, in a continuous manner while producing the final dosage forms at the end of the continuous step.

To implement HME from a batch mode to a fully continuous manufacturing process, the development and characterisation of the pharmaceutical products (produced either as powder or agglomerates), and subsequently its incorporation into final dosage forms (e.g., tablets or films) *via* a continuous mode of operation, has to be optimised primarily by two approaches: first, with an appropriate selection of automation, programming and controls set-ups (e.g., SIPAT, SCADA), micronising the agglomerated extrudates to granules and incorporating into final dosage forms; second, incorporating the extrudates into final dosage forms directly (**Figure 11.1**). This will be followed by in-line monitoring using at least two different PAT tools (NIR and Raman) to investigate critical quality attributes of the drugs used. The product quality attributes would mainly be monitored in different zones of the heated barrel of the extruder. Particle size distribution in the extruded materials would also be monitored *via* real time data acquisition using in-line particle size probe (Parsum, Germany) prior to and after the micronisation (by a rotor mill). This would then consequently be followed by a set-up of two gravimetric feeders/conveyor mixers in-line with the rotor mill. The first feeder would be used to add a controlled amount of additional excipients (such as lubricant or binder), required for the formulation of the final dosage forms to the second feeder and second feeder will be used to mix and convey the excipients with micronised extrudates (from the rotor mill) (**Figure 11.1**). The homogenously mixed powder formulations would be conveyed from the second conveyor mixer into a tablet machine for the continuous production of the tablets or into an injection moulding machine for the manufacture of implants. The whole manufacturing platform can be controlled and automated by using commercial software (e.g., SIPAT or SCADA) to integrate all the PAT tools that are necessary for continuous manufacturing into one overall platform. This will aim at a synchronisation between the feed-forward and feed-backward controls over the different unit operations in the whole line. On-the-line hardware automation and integration into a system to track the material mass flows over the complete manufacturing line then can be achieved.

## 11.5 Scale-up Methodologies

Process scale-up is necessary in order to produce large-scale manufacturing while maintaining the critical quality attributes of the product. A thorough investigation of the scale-up processes in continuous processing platform using HME can be optimised using a number of models and theories. However, the balance between the barrel, screw and die geometry and the melt temperature may not be constant before and after the scale-up. For continuous HME, scale-up can be achieved by running the extrusion for a relatively longer time or by increasing the material feed-rate. Also, increasing the barrel size and screw diameter or adding parallel units or transferring the heat across the barrel are considered some other ways of achieving a successful scale-up in pharmaceutical manufacturing. There may be some challenges involved in terms of the quality associated with each type of scale-up process, such as increasing the barrel size without changing the feeding rate, which may have different effects on the product quality. Moreover, changes in both the screw speed and the feed-rate may increase the shear force and torque inside the barrel, halting the process with significant impact on both the process and product quality. Overall, despite the difficulties and challenges, the scale-up *via* HME process can be optimised and undertaken by a very careful evaluation of process parameters and throughputs during the development stage.

The key to implementing a continuous manufacturing process *via* HME lies in the successful scale-up of the operation. Scale-up from a mid-size to a larger size twin-screw extruder is actually much more predictable, and the intense mixing that is normally achieved *via* an extrusion processing is more repeatable than batch mixing. This is only because a typical twin-screw extruder benefits from a shorter mass transfer which is inherent in its design. For a successful design of the scale-up involved in HME processing, there are several parameters that need to be managed, including the RTD using screw design, screw speed (rpm), and degree of screw fill. This should be coupled with a thorough understanding about the boundary conditions and process parameters and how these parameters in the event of a small size extrusion batch at the smaller scale, can be translated in terms of the quality attributes of the product.

Technically, for a successful scale-up of the HME twin-screw extrusion process, the fundamental geometry of the bench top extruder has to match that of the larger extruder. In this context the ratio of the outer diameter (OD) to the inner diameter (ID) of the screw is considered a key parameter (**Figure 11.2**). The screw elements of an extruder (twin-screw extruder) consist of mainly two types of screw blocks: (i) conveying blocks and (ii) mixing blocks. The first set of blocks are used to convey the materials through the barrel into mixing zones where the mixing blocks are used to perform intense mixing of the extrudates inside the barrel. Generally, in a typical twin-screw extruder, two types of mixing are observed – dispersive mixing and distributive mixing. Moreover, screws used in a twin-screw extruder are modular and can be easily changed to optimise the screw profile. In scale-up, the screw profile should be kept similar between the small twin-screw extruder and the large twin-screw extruder.

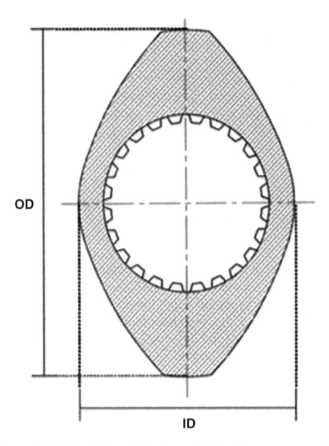

**Figure 11.2** The OD to ID of a typical twin-screw extruder

Mass or heat-transfer limitations can arise at larger-scales, thereby affecting dispersion, distribution and resulting product uniformity. In addition, with the increase on the screw diameter, the top speed of the rotating screw flights increases, thereby raising the peak shear to which the material is exposed. The effects of high peak shear can be reduced with tight acceptance windows and a specialised mixing element design which can potentially improve the self-wiping characteristics of a twin-screw extruder.

Design of experiment approaches alongside process models can prove useful methods for defining and predicting the process based design space. In addition, software tools can be used to predict and determine optimum scale-up conditions. However, there is no such software solution available on the market. The main reason amongst others is the requirement of raw material data, which can not only be difficult to calculate but also in some cases not feasible. In twin-screw extrusion processing, the mixture of components changes as it convey through the heated barrel under controlled conditions and converts from a solid to a non-Newtonian fluid. Research endeavours for the development of a robust software tool are on-going and may result in the creation of a complete software solution in the near future. Despite the innovative software revolution in scale-up *via* a twin-screw extruder, skilled personnel remain key to successful scale-up implementation in the real world.

Feeders (*via* which materials are fed into the extruder) are considered to be upstream processing equipment and should be taken into account during scale-up. In the current pharmaceutical extrusion paradigm there are two types of feeding process which are adopted, (i) split-feeding – the polymeric excipient and the active pharmaceutical ingredient(s) (API) are fed separately into the extruder, and (ii) premix feeding – a premix of the starting materials (polymer and API) is performed in a dry blend and subsequently then fed to the extruder in one stream.

It has been seen that in the commercial extrusion sphere, two types of feeding process are frequently used. Obviously, it depends on the applications as well as the physical forms of the feeding materials. A split-feeding process is a must in the event of feeding different physical forms (e.g., if excipient is pellets and API in fine powder). Since, premixing performs a great percentage of the mixing prior to the extrusion, it still outperforms split-feeding in terms of intense mixing because in a split-fed system the extruder does 100% of the mixing. The latter may lead to decreased rates or less mixing of the excipients with API. In comparison, a split-feeding system may prove to be only resort for processing API that is thermosensitive, requiring the feeding to happen somewhere in the middle of the extruder barrel after the excipient is at least partly melted. In laboratories, premix is still a preferred technique to split-feeding because small volumes (which are common in lab-scale) require specialised feeding equipment to meter and estimate the feeding accurately. Therefore, matching the feeding method during a successful scale-up can be very crucial as the feeding method significantly affects mixing and thus the quality attributes of the products. Similar to upstream equipment, downstream processes such as cooling, pelletising and micronising should be considered as part of the scale-up although these are frequently neglected in development timelines. The inconsistency or incomplete process with respect to downstream processing may lead to a tailback in the overall manufacturing process.

## 11.6 Regulatory Aspects

The FDA [21 Code of Federal Regulations (CFR) parts 210.3] defines the batch as, 'A specific quantity of a drug or other material that is intended to have uniform character and quality, within specified limits, and is produced according to a single manufacturing order during the same cycle of manufacture' [17].

Moreover, a lot is defined as 'a batch, or a specific identified portion of a batch, that has uniform character and quality within specified limits; or, in the case of a drug product produced by a continuous process, it is a specific identified amount produced in a unit of time or quantity in a manner that assures its having uniform character and quality within specified limits'. FDA (21 CFR 210.3) [17] allows flexibility in the definition of a batch. The underlying regulatory expectation of continuous manufacturing is promoted by the fact that the batch is of 'uniform character and quality within specified limits'. Being a 'batch' for the sake of quality assurance and the performance of the products, the definition itself is already in place to support the concept of a period of time. This interpretation itself will indeed assist and motivate pharmaceutical companies to move to continuous processing which is technically a single cycle of a manufacturing process.

Even though the overall issue of the introduction of new technology into the pharmaceutical manufacturing area has been very restrained by the FDA, the FDA has however recently issued a draft guidance [11] to the pharmaceutical industry in its '*Guidance for Industry PAT: A Framework for Innovative Pharmaceutical Manufacturing and Quality Assurance*'. The main aim and the goal of this guidance are to describe a regulatory framework within which industry and government can work together in order to increase the number and level of innovative pharmaceutical manufacturing technologies. The draft guidance document [11] states, 'PAT should help manufacturers develop and implement new efficient tools for use during pharmaceutical development, manufacturing, and quality assurance while maintaining or improving the current level of product quality assurance'. The guidance is regarded as central to the concept that while conventional pharmaceutical manufacturing is heavily involved with using batch mode, new opportunities exist to improve the efficiency and quality of the pharmaceutical manufacturing process. This is an attempt to introduce 21st century technology into the pharmaceutical industry in order to better respond to the rapidly changing marketplace for ethical pharmaceutical products. The draft guidance [11] also states, 'facilitating continuous processing to improve efficiency and manage variability'. Therefore, these regulatory statements will further encourage pharmaceutical industries to evolve from the frozen batch manufacturing to begin to exploit the benefits of continuous processing. Likewise, the regulatory authorities across the globe are emphasising the adoption of a continuous manufacturing process. These ideas are highlighted in the International Conference on Harmonisation guidelines especially in Q8 (R2), Q9, Q10, Q11 [18–21] and the FDA guidelines [11, 22]. Also the model describing the QbD approach encourages the continuous manufacturing in pharmaceutical industries. HME can be successfully adopted from a continuous manufacturing viewpoint by using these guidelines and the QbD model.

## 11.7 Summary

From a macroscopic viewpoint, it can be envisaged that a fully continuous HME process is possible to optimise and implement in a pharmaceutical sphere. A proper implementation of the continuous manufacturing process may reduce variations associated with product quality attributes and processing parameters. Thus, the ability to maintain the parameters and processing conditions may be the key to achieving operational benefits for continuous manufacturing *via* an emerging manufacturing platform (e.g., HME) and the subsequent scale-up for commercial exploitation. Recent guidelines from the regulatory aspects clearly envision an upcoming revolution of continuous manufacturing process over the current batch manufacturing pattern.

## References

1. S. Shah, S. Maddinenni, J. Ju and M. Repka, *International Journal of Pharmaceutics*, 2013, **453**, 1, 233.

2. J. Breitenbach, *American Journal of Drug Delivery*, 2006, **4**, 2, 61.

3. C.M. Crowley, F. Zhang, M.A. Repka, S. Thumma, S.B. Upadhye, S.K. Battu, J.W. McGinity and C. Martin, *Drug Development and Industrial Pharmacy*, 2007, **33**, 909.

4. M.A. Repka, S.K. Battu, S.B. Upadhye, S. Thumma, C.M. Crowley, F. Zhang, C. Martin and J.W. McGinity, *Drug Development and Industrial Pharmacy*, 2007, **33**, 1043.

5. C. Rawendaal in *Polymer Extrusion*, 4th Edition, *Hanser Publications*, Cincinnati, OH, USA, 2001, p.791.

6. R. Guy in *Extrusion Cooking: Technologies and Applications*, Woodhead Publishing, Cambridge, UK, 2001.

7. *International Symposium on Continuous Manufacturing*, 20–21st May, Cambridge, MA, USA, 2014, White Paper No.2,

8. P. Hurter, H. Thomas, D. Nadig, D. Embiata-Smith and A. Paone, *AAPS Newsmagazine: Manufacturing Science and Engineering*, 2013, **16**, 14.

9.  X.M. Zhang, L.F. Feng, S. Hoppe and G.H. Hu, *Polymer Engineering and Science*, 2008, **48**, 19.

10. J. Kossik in *Think Small: Pharmaceutical Facility Could Boost Capacity and Slash Costs by Trading in Certain Batch Operations for Continuous Versions*, Pharmamag.com, Article ID/DDAS-SEX 52B/. *http://www.pharmamanufacturing.com* [Accessed January 2015]

11. *Guidance for Industry PAT: A Framework for Innovative Pharmaceutical Manufacturing and Quality Assurance*, Center for Drug Evaluation and Research, US Food and Drug Administration, Rockville, MD, USA, August 2003. *http://www.fda.gov/downloads/Drugs/Guidances/ucm070305.pdf* [Accessed January 2015]

12. *Factory Shift: New Prescription for Drug Makers: Update the Plants*, Wall Street Journal, 3rd September 2003.

13. J. Dimasi, R. Hansen and H. Grabowski, *Journal of Health Economics*, 2003, **22**, 151.

14. D.B. Todd, *Polymer Engineering and Science*, 1975, **15**, 437.

15. G. Ganjyal and M. Hanna, *Journal of Food Science*, 2002, **67**, 6, 1996.

16. E. Reitz, H. Podhaisky, D. Ely and M. Thommes, *European Journal of Pharmaceutics and Biopharmaceutics*, 2013, **85**, 1200.

17. *21 CFR, Parts 210 and 211: Current Good Manufacturing Practice for Manufacturing, Processing, Packing, or Holding of Drugs*, Center for Drug Evaluation and Research, US Food and Drug Administration, Rockville, MD, USA. *http://www.fda.gov/cder/dmpq/cgmpregs.htm*

18. *ICH Q8 (R2) Pharmaceutical Development*, International Conference on Harmonisation, Geneva, Switzerland, 2009. *http://www.ich.org/fileadmin/Public_Web_Site/ICH_Products/Guidelines/Quality/Q8_R1/Step4/Q8_R2_Guideline.pdf*

19. *ICH Q9 Quality Risk Management*, International Conference on Harmonisation, Geneva, Switzerland, 2005. *http://www.ich.org/fileadmin/Public_Web_Site/ICH_Products/Guidelines/Quality/Q9/Step4/Q9_Guideline.pdf*

20. *ICH Q10 Pharmaceutical Quality Systems*, International Conference on Harmonisation, Geneva, Switzerland, 2008.
*http://www.ich.org/fileadmin/Public_Web_Site/ICH_Products/Guidelines/Quality/Q10/Step4/Q10_Guideline.pdf*

21. *ICH Q11 Development and Manufacture of Drug Substances (Chemical Entities and Biotechnological/Biological Entities)*, International Conference on Harmonisation, Geneva, Switzerland, 2012.
*http://www.ich.org/fileadmin/Public_Web_Site/ICH_Products/Guidelines/Quality/Q11/Q11_Step_4.pdf*

22. FDA Guidance for Industry: PAT: A Framework for Innovative Pharmaceutical Development, Manufacturing, and Quality Assurance.
*http://www.fda.gov/downloads/drugs/guidancecomplianceregulatoryinformation/guidances/ucm070305.pdf*

# Abbreviations

| | |
|---|---|
| AC | Anhydrous co-crystal |
| API | Active pharmaceutical ingredient(s) |
| AQOAT | Hydroxypropyl methylcellulose grade polymer |
| ASTM | American Society for Testing and Materials |
| CA | Citric acid anhydrous |
| CBZ | Carbamazepine |
| CFR | Code of Federal Regulations |
| CGMP | Current good manufacturing practice |
| CM | Citric acid monohydrate |
| Co-HME | Hot-melt co-extrusion |
| CP/MAS | Cross-polarisation/magic-angle spinning |
| CPS | Counts per second |
| CQA | Critical quality attribute(s) |
| CR | Controlled-release |
| CTZ | Cetirizine hydrochloric acid |
| DI | Discrimination index |
| DIF | Diffraction index data |
| DMF | Drug master file |
| DoE | Design of experiment(s) |
| DPD | Diphenhydramine hydrochloric acid |
| DSC | Differential scanning calorimetry |
| EC | Ethyl cellulose |
| EDX | Energy dispersive X-ray |
| EF | Extruded formulation(s) |
| EPO | Eudragit® EPO |
| ERC-SOPS | Engineering Research Center for Structured Organic Particulate Synthesis |
| EVAC | Poly(ethylene-*co*-vinyl acetate) |
| EXT | Extrudate(s) |

| | |
|---|---|
| FDA | US Food and Drug Administration |
| FI | Fragility index |
| FMT | Famotidine (BCS IV) |
| FT-IR | Fourier-Transform infrared |
| GMP | Good manufacturing practice |
| HC | Hydrated co-crystal |
| HIV | Human immunodeficiency virus |
| HM | Hot melt |
| HME | Hot-melt extrusion |
| HPC | Hydroxypropyl cellulose |
| HPLC | High-performance liquid chromatography |
| HPMC | Hydroxypropyl methycellulose |
| HPMCAS | Hydroxypropyl methylcellulose acetate succinate |
| HSM | Hot stage microscopy |
| HTE | High temperature extrusion |
| IBU | Ibuprofen |
| ICH | International Conference on Harmonisation |
| ID | Inner diameter |
| INM | Indomethacin |
| L/D | Length/diameter |
| MAS | Magnesium aluminometasilicate |
| MDX | Maltodextrin |
| MEPOL | Methacrylate hydrophilic polymer |
| MPC | Model predictive control |
| MSC | Multiplicative scatter correction |
| NCT | Nicotinamide |
| NIR | Near-infrared |
| NMR | Nuclear magnetic resonance |
| OD | Outer diameter |
| ODT | Orally disintegrating tablet(s) |
| PAT | Process analytical technology |
| PCA | Principal component analysis |
| PEG | Polyethylene glycol(s) |
| PEO | Polyethylene oxide(s) |
| PID | Proportional–Integral–Derivative |

| PLS | Partial least square(s) |
| PM | Physical mixture(s) |
| PMOL | Paracetamol |
| PRP | Propranolol hydrochloric acid |
| PRX | Piroxicam |
| PSD | Position sensitive detector |
| PVP | Polyvinyl pyrrolidone |
| QbD | Quality-by-design |
| QM | Quantum mechanical |
| RH | Relative humidity |
| RTD | Residence time distribution |
| SAC | Saccharin |
| SCADA | Supervisory control and data acquisition |
| SD | Standard deviation |
| SE | Specific energy |
| SEM | Scanning electron microscopy |
| SH | Steroid hormone |
| SIPAT | Siemen's PAT |
| SMEC | Specific mechanical energy consumption |
| SNV | Standard normal variate |
| SOL | Soluplus® |
| SR | Sustained-release |
| SSNMR | Solid-state nuclear magnetic resonance |
| TA | Theophylline anhydrous |
| TC | Theophylline-citric acid |
| $T_g$ | Glass transition temperature |
| $T_m$ | Melting temperature(s) |
| TSE | Twin-screw extrusion |
| VA | Vinyl acetate |
| VA64 | Kollidon® VA 64 |
| VP | N-vinylpyrrolidone |
| VRP | Verapamil hydrochloric acid |
| VSFL | Volume-specific feed load |
| VTXRPD | Variable temperature X-ray powder diffraction |
| WAXD | Wide-angle X-ray diffraction |

| | |
|---|---|
| WAXS | Wide-angle X-ray scattering |
| XPS | X-ray photo-electron spectroscopy |
| XRPD | X-ray powder diffraction |

# Index

## A

Absorb, 126
Absorbance, 59, 66, 164
Absorption, 122, 171
Acceleration, 80, 85, 89-90
Acceptor, 100, 127
Acid, 6, 59, 64-65, 76-78, 86-91, 99-101, 104, 124, 141
Acidic, 132, 146
Acryl-EZE®, 7, 102, 107-115
Activation energy, 158
Active pharmaceutical ingredient(s) (API), 5-6, 14, 19-21, 29, 31-33, 35-36, 41-44, 53, 56, 58, 65, 68-70, 75, 79, 87, 91-92, 97-102, 104-108, 110-111, 113, 115, 123-124, 129-132, 143, 149-150, 154, 159, 170-173, 175-176, 178, 191
Additive(s), 6, 29, 105, 129, 169, 171, 173-174, 178
Adhesion, 55-56
Adhesive, 106, 123
Adsorbent, 65
Agglomerated, 53, 183-184, 188
Agilent Technologies, 141
Agitation, 90, 149, 170
Air, 25, 32, 128, 169
  cooling, 169
  entrapment, 25
Alpha MOS, 99, 107
Aluminium, 123, 140, 153, 169
Ambient, 5, 68, 81, 87, 97, 153, 159-162
  temperature, 159, 162
American Society for Testing and Materials (ASTM), 46
Amide, 65-66, 99
Amine, 102, 155

Amorphous, 11, 20, 39, 42-43, 53, 58, 63-65, 68, 71, 79-81, 83-84, 87-88, 90, 92-93, 101-102, 106, 110-111, 124, 130-132, 139, 143-144, 150, 152, 158-159, 181
  phase, 64, 79-81, 83-84, 87-88, 150
Amorphicity, 59, 63-64, 131
Amorphisation, 64, 81
Ancillary equipment, 177
Angle geometry, 37
Anhydrous, 6, 86, 91-92
  co-crystal (AC), 86-87
Anionic, 99-102
Anode, 124, 141, 153
Antaris II NIR spectrometer, 152
Antioxidant(s), 6, 169
Anton Paar TTK450, 153
Aqueous, 20, 55, 63, 118, 139
Arrhenius law, 80
Ascorbic acid, 6
Assay, 124, 141
Astree e-tongue, 99, 101, 107-108, 112, 114
  data, 112
At-line monitoring, 36, 46
Atomic force microscopy, 36

## B

B3LYP 6-31G, 102
Barrel, 3, 9-11, 13, 21-25, 53-55, 69, 83, 90, 105, 169-170, 172, 175-176, 184-185, 187-191
  liner, 175
BASF polymers, 151
Batch manufacture(ing), 8, 28, 35, 38-39, 44, 121, 185, 187, 192-193
Batch mode, 185, 187-188, 192
Batch process(es), 19-20, 27, 29, 35, 45, 181-182, 187
Batch size, 38, 187
BAY 12-9566, 65
Bi-lobal mixing, 128
Binary, 53, 104, 110, 130, 144, 152, 155, 160
Binder, 188
Binding, 102, 126-127
  energy, 102, 126-127
Bioadhesive, 106
Bioavailability, 1-2, 4-5, 8, 20, 55, 63, 68, 75, 79, 122, 149, 181

Biodegradable, 6, 9, 57
Biological, 46, 122, 195
Biopharmaceutical, 8, 68
Bitter, 53, 97-101, 104-107, 109, 111, 113-115, 117, 119, 170
Bitterness, 98-100, 107, 114
Blend, 25, 31, 36-37, 54, 58, 140, 152, 191
Blended, 91, 187
Blending, 31, 35-36, 44, 105, 184, 187-188
Bond, 94, 99, 102, 127
Bonding, 56, 64, 66, 69, 98, 100, 102, 109, 151
Bruker, 124, 141, 153
Buccal cavity, 104
Bulk, 23, 141-143, 145, 156-157, 159-160
    density, 23

## C

Caffeine, 76, 90
    -oxalic acid, 76, 90
Calender, 179
Calendering, 7, 21, 106
    equipment, 21
Calibration, 32, 141
Calorimeter, 123, 153
Calorimetry, 4, 58, 66, 85, 91, 100, 130, 140, 142, 153, 165, 167, 171, 184
Carbamazepine (CBZ), 70, 76, 78-79, 81-85, 89-93
Carbon dioxide ($CO_2$), 171, 175, 178
Carbon footprint, 181
Carboxyl group, 99, 155
Carrier, 6, 11, 20, 54, 64-65, 70, 92-93, 97, 100, 105-106, 124, 140, 144, 153, 183
    matrix, 97
Cationic, 58, 99-100, 102
Cavity, 23, 104, 106
Cellulose, 6-7, 43, 104, 106, 121
    acetate, 7
Cetirizine hydrochloric acid (CTZ), 101-102, 106-115
Characterisation, 33, 36, 45, 69-70, 100, 109, 122, 124, 128, 131, 140, 150, 153, 156, 188
Charged, 8, 98, 121
Chemical, 6, 16, 27, 29, 37, 40, 44, 46-47, 50-51, 56, 60, 69, 75-76, 78, 83, 95, 98-99, 102, 116, 123, 147, 149, 151, 154, 165, 185, 195
    composition, 123

industry, 60
shift, 102
stability, 6
Chemistry, 17-18, 37, 44, 47, 49, 51-52, 77, 88, 137, 139, 148, 166
Chromatography, 124, 140-141
Chrono-pharmaceutical dosage, 2
Cinnamic-acid, 78, 89
Citric acid anhydrous (CA), 86-87, 107
Citric acid monohydrate (CM), 59, 66, 86-87, 100, 153, 163-164
Clean, 175, 177, 179
Cleaning, 171, 175-178
Clinical performance, 8, 68
Clotrimazole, 106
Co-crystal, 8-9, 68, 75-93
  formation, 9, 80-81, 83, 85-86, 88-91
Co-crystallisation, 68, 75, 77-81, 83, 85-89, 91-93, 95
Co-evaporation, 139, 150
Co-extrudate, 55
Co-extruded, 54-55, 57, 146
Co-extrusion, 41-42, 53-59, 61, 139
Co-milling, 139
Co-precipitation, 150
Co-rotating, 3, 9, 12, 18, 25, 69, 176, 178
  extruder, 18, 25
Coat, 32, 55-56, 58
Coated, 22, 65, 112, 123-124, 132-135
Coating, 7, 28, 32, 35, 41, 98, 102, 104-106, 114, 132-135, 188
Code of Ethics, 107
Code of Federal Regulations (CFR), 30, 45-46, 49, 191-192, 194
Colorcon Ltd., 107
Colorimeter, 12
Colorimetric, 12
Colour, 12, 31
  concentration, 12
  pigment, 12
Compaction, 7, 19, 39, 187
COMPASS software, 123
Compliance, 7, 27-28, 35, 53, 55, 57, 59, 61, 97-99, 175
Composition, 26, 31, 111, 123, 131-133, 144, 159
Compound, 6, 56, 63, 69, 172-173
Compounding, 18, 127
Compressed, 23, 32, 37, 67, 100, 128, 132

Compressibility, 4, 8, 63, 68, 76
Compressible, 76
Compression, 3, 23-24, 33, 41, 149, 187
  and filtration properties, 149
  zone, 3, 23-24
Concentration, 12, 45, 64, 67, 91-92, 99, 108
Consumption, 11, 185
Contamination, 176-178
Continuous extrusion process, 22, 154
Continuous manufacturing, 1-31, 34-42, 44-62, 64, 66, 68, 70, 72, 76, 78, 80, 82,
    84, 86, 88, 90, 92, 94, 98, 100, 102, 104-106, 108, 110, 112, 114, 116, 118,
    121-137, 139-140, 142, 144, 146, 148, 150, 152, 154, 156, 158, 160, 162,
    164, 166, 170, 172, 174, 176, 178, 180-195
Continuous process, 20, 30, 42, 44, 63, 65, 67, 69, 71, 73, 122, 170, 184-185,
    188, 192
Control, 1-2, 5, 9, 19-21, 24, 27-28, 30-31, 34-35, 39-41, 44-46, 51, 57, 75, 78,
    90, 122, 128, 132, 149, 170, 176-179, 182-185, 187
Controllability, 27, 182
Controllable temperature, 79-80
Controlled, 1, 3, 7, 15-16, 21, 28, 33, 35, 40, 53, 97, 117, 124, 132, 136-137,
    141, 153, 170, 180, 182, 188, 191
  -release (CR), 15-16, 53, 58, 117, 136-137, 180
Controlling, 1, 3, 45, 121, 133, 171
Conversion, 65, 70, 75, 78-81, 83-91, 93, 110, 150
  process, 81, 84
Converting, 186
Conveyor, 3, 37, 128, 177, 188
  belt, 37, 128, 177
  mixer, 188
Cooled, 169
Cooling, 2, 4, 20-22, 26, 37, 41, 75, 142, 149, 169, 171, 175, 178-179, 191
  crystallisation, 75
  rate, 21, 26, 149
  system, 178
  zone, 37
Counter-rotating, 9, 25
  extruder, 25
Counts per second (CPS), 103
Covalent, 69, 75, 159
Critical quality attribute (CQA), 36
Crowell, 134
  Hixson-Crowell, 124-125, 134-135

Cross-polarisation/magic-angle spinning (CP/MAS), 86
Crosslinked, 101
Cryogenic grinding, 81
Cryomilling, 42
Crystal, 8-9, 63, 68, 72-73, 75-95, 149-151, 157, 159
    transformation, 159
Crystalline, 43, 58-59, 63-65, 71, 75, 106, 110-111, 124, 130-132, 144, 149-150,
        152, 154, 156-159, 162
    peak, 159
    structure, 71, 156
Crystallinity, 64, 144, 157-158
Crystallisation, 9, 20, 33, 44, 68-69, 75-81, 83, 85-89, 91-93, 95, 149-150
    process, 9, 79-81, 83, 149
Crystallised, 93
Cubic law, 10-11
Current good manufacturing practice (CGMP), 45-46, 194
Cutting, 3-4, 44, 179
Cyclical, 23
Cyclodextrin inclusion complex, 42


**D**

D-gluconolactone, 70
D8 Advance, 124, 141, 153
Damage, 123
Decay, 22
Declaration of Helsinki, 107
Degradation, 5-7, 13, 31, 92-93, 171, 173-175
    point, 173-174
Degree of crystallinity, 157-158
Degree of polymerisation, 152
Deionised, 107-108, 113
    water, 107-108
Delamination, 57
Delivery rate, 23
Delivery system, 20, 29, 37, 57
Deprotonation, 100
Design of experiment(s) (DoE), 12, 35, 38, 127, 190
Detector, 124, 141, 152-153
Devolatilising, 24
Diagnostic Instruments, Inc., 124, 141, 153
Die(s), 1-3, 12, 21-24, 26, 31, 33, 37-38, 41-42, 53-56, 58-59, 104, 115, 122,
        128, 140, 147, 152, 169, 184, 188

cavity, 23
design, 37
diameter, 104, 115
geometry, 37, 188
resistance, 24
size, 26, 37
swell, 55
Differential scanning calorimeter, 123, 153
Differential scanning calorimetry (DSC), 4, 58-59, 66, 68, 70, 85, 89, 91, 100, 109-111, 130-131, 140, 142-144, 153-154, 157-159, 171, 184
transition, 131
DiffracPlus Commander, 153
Diffraction, 36, 58, 66, 68, 70, 81, 84, 87, 91, 100, 124, 141, 144, 150, 156, 161-162, 166, 184
index data (DIF), 161
Diffractogram, 132, 144, 156
Diffusion, 56, 80, 87, 125, 133, 135
Dimer, 59, 66
peak, 59
Diphenhydramine hydrochloric acid (DPD), 100-102
Discrete Element Model, 35
Discrimination index (DI), 112-114, 151
Dispersion, 1, 7, 20, 37, 42-43, 58, 65-66, 69, 93, 100, 102, 109, 112, 131-132, 150-151, 172, 181, 190
Dispersive mixing, 38, 90, 189
Dissolution, 5, 10, 20, 35, 41-42, 53, 58, 63-68, 70, 75-76, 78-79, 91-93, 100-101, 105-106, 111, 113-114, 124-125, 131-134, 139-141, 145-146, 149-150, 181, 184
mechanism, 124
rate, 5, 10, 20, 35, 42, 53, 63-64, 66-68, 70, 91-92, 100, 105, 125, 134, 139, 149, 181
Dissolved fraction, 125-126
Dissolving, 8, 68, 106, 113, 133
Distributive mixing, 127, 189
Dosage, 1-2, 4-7, 9, 14, 19-20, 26-30, 32, 35, 37, 42, 54-55, 58, 60, 69, 97-99, 105-106, 118, 121-122, 125-126, 128, 170, 181, 185, 187-188
form, 2, 5-6, 27-30, 32, 37, 42, 122, 125-126, 170, 185, 187
formulation, 20
Dose, 1, 7, 40, 57, 108, 121
dumping, 1, 7, 121
Dosing, 7, 170-171
Downstream equipment, 175, 177-179

Downstream process(ing), 2, 21, 56, 121, 169, 177-178, 184, 191
Drug, 1-2, 5-9, 14-20, 26-29, 33, 35-39, 45, 47-73, 77, 92-93, 97-102, 104-106,
    108-112, 114-119, 121-137, 139-147, 149-152, 154-159, 162, 165-166, 170-
    171, 176, 179-181, 183, 186-187, 191-195
  absorption, 122, 171
  artemisinin, 70
  complexation, 98
  delivery, 7-9, 14, 20, 45, 50, 52-54, 57, 60-62, 69, 97, 115, 118, 129, 150, 165,
    180-181, 193
  dosage form, 126
  loading, 66-67, 102, 111, 126, 162
  master file (DMF), 29
  molecule, 126-127
  -polymer, 20, 28, 36-37, 67, 99-102, 109, 111, 114, 126-127, 129, 131, 135,
    144, 151
  ratio, 37-38, 70
  release, 7, 28, 54, 57-58, 66, 70, 104, 106, 122, 124-126, 133-135, 141, 146
    rate, 58, 122, 134
Dry, 19, 39, 43-44, 53, 77, 105, 140, 153, 187, 191
  blend, 191
  granulation, 19, 39, 77, 187
Drier, 33
Drying, 1, 19, 21, 28, 33, 35, 44, 54-55, 69, 98, 139, 150, 188
Dynamic(s), 20, 32, 40-41, 79
Dynasan® 114, 105

**E**

e-tongue, 99, 101-102, 107-108, 112-115
Easy clean system, 175, 177, 179
Efficacy, 2, 57, 68, 115
Efficiency, 5, 19, 27, 67-68, 87, 100-101, 104-105, 111, 113-114, 176, 192
Electric, 22, 24
Electrical, 21, 178
Electron, 58, 65, 99, 109, 123, 156
  beam, 123
Electronic, 49, 99, 101, 106, 126
  energy, 126
  -tongue, 101, 106
Encapsulation, 56
Endotherm, 67, 100, 110, 143, 157
Endothermic, 59, 66, 110, 130, 142, 157
  peak, 130

transition, 130, 157

Energy, 5, 11, 13, 20, 24, 27, 40, 56, 59, 69, 79-81, 102, 108, 123, 126-127, 151,
    154, 158, 181, 185, 187
    consumption, 11, 185
    dispersive X-ray (EDX), 59, 123, 129

Engineering, 8-9, 16, 18, 20, 31, 44, 47, 50-51, 61, 63, 65, 67-69, 71, 73, 98, 116,
    137, 166, 193-194

Engineering Research Center for Structured Organic Particulate Synthesis
    (ERC-SOPS), 20

Enrofloxacin, 104

Enthalpy, 70, 91, 110, 157-158

Entropy, 24, 158

Environment, 150, 169, 171, 176-177, 185

Environmental, 46, 75, 105, 184

Equilibrium, 69, 158

Equimolar, 79-80, 85, 91

Ethanol, 85-87, 89

Ethics Committee of the University of Greenwich, 107

Ethyl cellulose (EC), 7, 121-122, 126, 130-134
    N10, 122, 126, 130-134
    P7, 122, 126, 130-134

Ethylene, 6, 57
    vinyl acetate, 57

Eudragit®, 6-7, 65, 99-102, 106-115, 117
    EPO (EPO), 65, 67, 100-102, 104
    L, 7
    L100, 99, 101-102, 107-115
    L100-55, 99, 101-102, 107, 110-111, 114
    RSPM, 7
    S100, 102

Eurolab-16, 58, 122, 140, 152

European Medicines Agency, 27, 57, 62

Eutectic, 65, 70, 76, 79-80, 83-84, 88-89, 149
    mixture, 65, 149
    peak, 80
    point, 70, 76, 80, 83, 88-89

EVA V.16 software, 153

Evaporation, 1, 42, 54, 75, 123, 139, 150

Evonik Pharma Polymers, 107

Exothermic, 130, 143

Extrapolation, 158

Extrudate(s) (EXT), 1, 6, 20-21, 23-24, 26, 31, 36-38, 55, 109, 129, 140-146,
    152-153, 156-161, 163-164

Extrude, 55, 100
Extruded, 1, 12, 21-22, 36-37, 41, 54-55, 57, 59, 66-70, 76, 80, 84-85, 88-93, 99-102, 104-106, 108-109, 111-112, 121, 126-129, 131-132, 135, 141-146, 150-153, 155, 158-159, 184, 188
  formulation(s) (EF), 55, 99-102, 107, 109-112, 114-115, 129-133, 144, 150, 156-160, 162-163
Extruder, 2-4, 9-13, 18, 21-22, 24-25, 37, 58, 69-70, 76, 83, 86-88, 105, 122, 128, 140, 152, 170-179, 184-185, 187-191
  die, 12
  screw, 24, 105, 128
Extrusion, 1-31, 34-42, 44-73, 76-80, 82, 84, 86-92, 94, 97-100, 102, 104-106, 108, 110, 112, 114, 116-118, 121-137, 139-167, 169-195
  process(ing), 1-3, 6, 10-11, 14, 18, 22, 24, 31, 36, 55, 67, 69, 71, 92, 97, 99-100, 104-106, 122, 127, 139-145, 147, 149-155, 157, 159, 161-163, 165, 167, 175, 177-178,181, 184, 186, 189-190
  system, 177
  temperature, 7, 37-38, 55, 160
ExtruVis 2, 12

**F**

Famotidine (BCS IV) (FMT), 68
Fed, 22, 37, 128, 185, 191
Feed, 1, 3, 5, 12-13, 21-24, 31, 34, 38, 41-42, 58, 76, 91, 93, 127-128, 140, 142, 152, 176-177, 184, 187-189
  -rate, 1, 12-13, 31, 58, 76, 91, 93, 127-128, 140, 142, 152, 184, 187, 189
  section, 22-24
  zone, 3
Feeder, 37, 42, 128, 140, 169, 183, 188
Feeding, 2, 9, 12, 21-24, 31, 35, 37, 58, 105, 122, 152, 182-184, 189, 191
  method, 191
  process, 191
  rate, 9, 189
  section, 12, 22-23
  zone, 37, 58, 122, 152
Feedstock, 22-23
Fenofibrate, 43, 105
  NXP 1210, 105
Fibre, 128, 152-153
Fick's, 135
Fickian, 135
Filled, 37, 55
Filler, 176

Filling, 13
Film(s), 2, 26, 32, 98, 105-106, 121, 181, 188
    forming, 105-106
First order kinetics, 124-125, 133
First order release, 125-126
Fisher Chemicals, 107, 140
Flexibility, 6, 19, 27, 105, 185, 192
Flexible, 8, 19, 39-40, 57, 105, 121, 173
Flow, 2-3, 5, 21-23, 28, 35-39, 41-42, 44-45, 55, 70, 76, 85, 141, 176, 182,
    184-185
    function, 76
    rate, 41-42, 141
Flowability, 39, 76
Fluid, 20, 32-33, 35, 69, 139, 150, 191
Fluidised bed coating, 98, 106
Food, 1, 5, 17-18, 27, 47, 49, 52, 54, 72-73, 105, 121, 148, 169-170, 181,
    185, 194
    industry, 5, 54, 169, 181
Force, 3, 7, 21-23, 33, 36, 79-80, 84, 90-91, 127, 155, 189
Fourier-Transform infrared (FT-IR), 58-59, 91, 100
FP82HT hot stage, 124, 141, 153
FP 90 central processor, 124, 141, 153
Fraction, 41, 57, 70, 82, 125-126, 152, 157-158
Fragility index (FI), 158
Free energy, 56, 79, 81, 126
Free radical, 6
Free volume, 173
Freeze, 1, 33, 54, 98, 139, 150
    -drying, 1, 33, 54, 98, 139, 150
Freezing, 26
Friction, 22-24
Frozen, 20, 28-29, 185, 192
    batch manufacturing, 20, 28, 192
    batch mode processing, 185
    batch processing, 185
Fusion, 154, 157
    enthalpy, 157

## G

G-50, 140, 142-146
Gas(es), 152, 158, 178
Gastro-resistant, 7

Gastrointestinal, 20, 122
Gattefosse, 140
Gaussian-09, 102, 126
Gelucire, 65
 -50/13, 65
Geometric, 10, 126
Geometry, 3-4, 9, 22-23, 37, 69, 178, 188-189
Glass, 6, 26, 59, 66, 100, 130-131, 142-143, 150, 152, 157-159, 171
 slide, 131
 transition, 6, 26, 59, 66, 100, 130, 142-143, 150, 152, 157-158, 171
 temperature ($T_g$), 6, 26, 36, 59, 66-67, 100, 110, 130-131,142, 152, 154, 157-
 158, 162, 171
Glassy, 26, 67, 100, 158
 solution, 67, 100
 state, 26
Glycerol distearate, 104-105
Glycerol trimyristate, 105
Glyceryl dibehenate, 104
Glyceryl monostearate, 104
Glyceryl tripalmitate, 104
Goebel mirror, 124, 141, 153
Good laboratory practice, 29
Good manufacturing practice (GMP), 29, 45, 171, 175, 194
 compliance, 175
Gordon–Taylor equation, 110, 131
Granulation, 19, 35, 39, 43, 64-65, 77, 187-188
Granule, 31, 37
Grinding, 2, 75-76, 79, 81-85, 87

**H**

Hansen, 116, 142, 151, 154-155, 166, 194
 Solubility Parameter, 151, 154-155
Hardening, 6
Hardness, 41-42
Health, 9, 44, 49, 52, 105, 194
Heat, 5, 10-11, 20, 22, 85, 87, 90-93, 97, 154, 172, 189-190
 flow, 85
 transfer, 10-11
Heated, 3, 22, 24, 53-54, 105, 123, 140, 153, 169, 188, 190
Heating, 3, 21, 80, 84-85, 89, 93, 131, 143, 153, 158-159, 171, 187
 phase, 171
 process, 84-85, 93

    rate, 84, 89, 153, 158
Helix angle, 22, 37
High-performance liquid chromatography (HPLC), 107, 124, 140-141
High temperature, 70-71, 79-81, 83-84, 175
    extrusion (HTE), 70-71
Higuchi, 124-125, 134, 137
    model, 125, 134
Hixson, 134
    Hixson–Crowell, 124-125, 134-135
Hoftyzer, 108, 116, 126, 151, 154, 166
    Van Krevelen–Hoftyzer, 108, 126, 151, 154
Homogeneity, 26, 36, 92
Homogeneous, 2, 23, 37, 41, 123
Homogenises, 3
Homogenous, 64, 129, 144
Homogenously, 128-129, 188
Hopper, 2-3, 12, 21-22, 25, 37, 41, 105, 183
Horizontal processing, 176-179
Hot melt (HM), 159, 163
    co-extrusion (Co-HME), 53-54, 59, 139-140
    extruded, 105-106, 151
    extrusion (HME), 1-31, 33-43, 44-56, 58-60, 62-64, 66, 68-70, 72, 76, 78, 80,
        82, 84, 86, 88, 90, 92, 94, 97-102, 104-106, 108, 110-111, 112, 114-118,
        121-122, 124, 126-136, 139-167, 169-170, 172, 174-176, 178-195
    process(ing), 1, 2, 4, 7, 10-12, 14, 18, 20, 26, 36-37, 40, 45, 53-54, 64, 70,
        97-99, 104, 106, 110, 121-122, 129, 131, 139-143, 145, 147, 149-153, 155,
        157, 159, 161, 163, 165, 167, 181, 183-185, 187, 189, 193
Hot spin mixing, 139, 150
Hot stage microscopy (HSM), 36, 70-71, 124, 131, 140, 143-144, 153, 159
Human immunodeficiency virus (HIV), 43, 122
HIV-1, 122
Humidity, 64, 79, 81
HYCROME-4889, 141
Hydrated co-crystal (HC), 87
Hydrochloric acid, 99-101, 124
Hydrocortisone, 106
Hydrogen, 64-66, 69, 98-100, 102, 109, 126-127, 151
    bond(ing), 64-66, 69, 98, 100, 102, 109, 127, 151
    donor, 100
Hydrophilic, 5, 58, 64-65, 68, 70, 106, 132, 139
    carrier, 65, 70, 106
    polymer, 5, 58

Hydroxypropyl cellulose (HPC), 106
Hydroxypropyl methycellulose (HPMC), 7, 106, 121, 132, 140, 142-146
  acetate succinate (HPMCAS), 143
  grade polymer (AQOAT), 142
Hygroscopicity, 75
Hysteresis, 171

**I**

Ibuprofen (IBU), 7, 67, 76, 80, 89-90, 98, 100, 121
Immiscible, 26, 56, 108-109, 126, 154
Impact, 9-10, 20, 23-24, 35, 37, 68, 70, 169, 176, 184-185, 187, 189
Implanon®, 57
Implant, 26, 43
Implantable, 57
Impurity(ies), 22, 78, 118, 149, 185
*In situ*, 81, 84, 91-93
*In vitro*, 58, 64, 66-68, 92, 99, 101-102, 104, 106-108, 111-112, 115, 124,
    132-133, 141, 145-146
*In vivo*, 99-100, 102, 104, 106-107, 111, 115
In-line monitoring, 19, 36, 152, 183, 188
  test, 36
Indomethacin (INM), 63-65, 68, 141-142, 144, 146
Inert, 53, 98
Infrared, 4, 12, 21, 47, 58, 83, 91, 100, 127, 163, 183
  spectroscopy, 47, 163
InGaAs detector, 152
Injection, 5, 106, 141, 183, 188
  moulding, 5, 106, 183, 188
Inner diameter (ID), 189-190
Inorganic, 26, 53-54, 63-64, 182
  carrier, 64
Insight QE camera, 124, 141, 153
Insoluble, 8, 53, 68, 100, 139-140, 181
  drug, 140
Institute für Nanomaterialien in Saarbrücken, 174
Intensity peak, 132, 160
Inter-subject variability, 1, 121
Interaction, 37, 58-59, 64, 66, 70, 79, 92, 99-100, 102, 127, 131, 151-152,
    154-155, 159
  energies, 131, 159
  parameter, 151-152, 154-155
Interdiffusion, 56

Intermeshing, 25, 178

Intermolecular, 58-59, 65-66, 79-80, 90, 92, 98-100, 102, 108, 122, 126-127, 142-143, 154
  contact, 79, 90
  interaction, 58-59, 79, 92

International Conference on Harmonisation (ICH), 9, 17-18, 30, 72-73, 118, 192, 194-195

Intra-subject variability, 1, 121

Intramolecular, 102, 108, 126, 154

Ion, 6, 98
  exchange, 98
  Ionic, 100

**K**

Kaletra®, 7, 43

Ketoconazole, 106

Ketoprofen, 65

Kinetic(s), 20, 85, 124-125, 133

Kneading, 2, 22, 128, 183
  block, 128

Kollidon®, 68, 101-102, 154
  VA 64 (VA64), 68, 101, 154-155, 157, 162

Korsmeyer–Peppas model, 125, 135

**L**

Layer, 55-58, 98-99, 104, 132

Leakage, 23

Length/diameter (L/D), 3, 12, 24, 172, 178

Light, 6, 124, 141, 143, 153, 172-174

Linear, 133-135, 158, 173
  extrapolation, 158

Linearity, 133-135

Lipid, 97, 104-105, 139-144, 146
  -based cold extrusion, 105
  -based formulation, 139

Lipidic, 63, 65, 98, 104, 139, 144
  carrier, 65, 144
  matrices, 104

Liquid, 21-22, 26, 65, 75, 79-80, 82, 87, 124, 140-141, 149, 158, 169
  chromatography, 124, 140-141
  phase, 79
  state, 65, 149, 169

Load, 13, 56

Loading, 66-67, 99, 102, 111, 121, 126, 157-158, 162, 173

Low temperature, 81, 85-87, 157

Lubricant, 31, 41, 188

LYNXEYE™, 124, 141, 153

LynxIris, 124, 141, 153

## M

Macroscopic, 79, 193

Magnesium aluminometasilicate (MAS), 58-59

Magnesium aluminosilicate, 65

Magnetic, 83, 102, 108

Mallinckrodt Chemical Ltd., 151

Maltodextrin (MDX), 106

Manufacture, 2, 4, 9, 40, 42, 53, 64, 69, 75-76, 93, 98, 115, 121, 127-128, 150, 176, 181, 184, 187-188, 191, 195

Manufactured, 1, 5, 9-10, 12, 21, 26, 55, 64, 66, 104, 121-122, 128-129, 185

Manufacturing, 1-31, 33-42, 44-62, 64, 66, 68, 70, 72, 76, 78, 80, 82, 84, 86-88, 90, 92, 94, 98, 100, 102, 104-106, 108, 110, 112, 114, 116, 118, 121-137, 139-140, 142, 144, 146, 148, 150-152, 154, 156, 158, 160, 162, 164, 166, 170-172, 174, 176, 178, 180-195

   process, 1, 9, 11, 14, 20, 24, 28-30, 36, 39, 41, 45, 54, 70, 128, 139, 182, 184-189, 191-193

Market, 8, 27, 31, 39, 44, 54-55, 57, 63, 68, 175-176, 179, 186, 190

Mass transfer, 79-80, 189

Material, 1-6, 8, 11-13, 21-25, 28-29, 31, 35, 37-39, 41, 45-46, 48, 54, 56, 64, 68, 75, 81, 87, 121, 123, 131, 169, 172, 176, 185, 187-191

Matrix, 1, 5-7, 12, 26, 53, 55, 59, 66-67, 91, 97-100, 102, 104, 110, 114-115, 121, 125-126, 128-129, 131-135, 139, 142-144, 150, 157-159

   surface, 134

   tablet, 132

Mechanical, 11, 13, 26, 37, 56, 69-71, 75-77, 79, 81, 84, 106, 126, 149, 172-173, 185

   activation, 81

   energy, 11, 13, 185

   interlocking, 56

   method, 69

   properties, 70, 75-76, 149

Mechanochemical, 79, 81

   synthesis, 79

Medicines and Healthcare Products Regulatory Guidelines, 27

Melt, 1-31, 34-54, 56, 58, 60, 62-73, 76, 78, 80, 82, 84, 86, 88-90, 92, 94,
    97-100, 102, 104-106, 108, 110, 112, 114, 116-118, 121-137, 139-167,
    169-174, 176, 178, 180-195
    extrusion, 1-31, 34-42, 44-54, 56, 58, 60, 62-73, 76, 78, 80, 82, 84, 86, 88, 90,
        92, 94, 97-100, 102, 104-106, 108, 110, 112, 114, 116-118, 121-137,
        139-167, 169-170, 172, 174, 176, 178, 180-195
    flow process, 36-37
    method, 42, 53
    phase, 80
    pool, 24
    pressure, 171
    process, 10, 84
    temperature, 173-174, 188
    viscosity(ies), 21, 26, 37, 56
Melting, 2, 5-6, 10, 13, 22, 24, 26, 37, 41, 53-54, 59, 66-67, 75-76, 80, 84, 93,
    100, 104, 110, 130-131, 140, 142-143, 149-152, 154-155, 157-159, 170-171
    method, 75, 84
    peak, 59, 66, 110, 130, 142, 152, 157-158
    point, 5, 130, 140, 150-151, 158, 171
    state, 155
    temperature ($T_m$), 10, 12, 26, 37, 76, 80, 83, 89, 92-93, 130, 149, 152,
        154-155
    zone, 37
Meltrex®, 7
Metastable, 70, 150, 163
Metering zone, 3-4, 23
Methacrylate hydrophilic polymer (MEPOL), 58-59
Mettler–Toledo International Inc., 124, 141, 153
    -823e, 123, 140, 153
Microcrystalline, 104, 106
    cellulose, 104, 106
Microencapsulation, 98, 115-116, 175, 180
Micronised, 58, 188
Micronising, 188, 191
Microscope, 123-124, 140, 153
Microscopy, 36, 58, 65, 70, 109, 123-124, 140, 153, 156
Mill, 37, 41, 123, 183, 187-188
Milled, 104, 152, 187
Milling, 22, 42, 58, 64, 81-82, 109, 139, 152, 177, 184, 188
    equipment, 177
    process, 109
Miscibility, 26, 36, 67, 100, 109, 126, 142, 151, 154-155

Miscible, 56, 108-110, 126, 131, 142, 149, 154, 159
Mix, 66, 171, 188
Mixed, 23, 37, 58, 122, 140, 152, 169, 188
Mixer, 58, 122, 140, 152, 183, 187-188
Mixing, 1-3, 6, 9, 13, 22, 24-25, 37-38, 41, 54, 64, 69, 76, 80, 83, 88, 90-92, 98,
    108, 126-128, 131, 139, 144, 150, 154, 159, 163, 170, 172-174, 187,
    189-191
  speed, 1
  zone, 3, 37
Mixture, 1-2, 41, 65, 70, 83-85, 91-92, 105, 130, 149, 155, 190
Mobility, 65, 80-81, 90
Model predictive control (MPC), 28, 40-42, 45
Modified-release, 97, 118
Moisture, 4, 39, 97
Molar, 70, 151-152, 154
  attraction, 151
  dispersion, 151
  ratio, 70
  volume, 151-152
Molecular, 5-7, 16, 36-37, 55-56, 66, 68-70, 80-81, 86-87, 93, 95, 97, 100, 106,
    110, 126-127, 131, 135, 147, 150, 154-155, 166
  adhesion, 56
  bonding, 56
  contact, 81, 86
  diffusion, 80, 87
  dispersion, 66, 100
  mobility, 80-81
  modelling, 126-127, 131, 135
  weight, 5-7, 36, 106
  dispersed, 20, 57, 59, 67, 98, 100-101, 114-115, 139
Molten, 6, 22, 24, 54, 65, 124, 140, 153
  drug, 6
  mass, 22, 24
  polymeric carrier, 124, 153
Monitor, 4, 9, 128, 163
Monitoring, 3, 8, 19, 21, 27-36, 39, 42, 44-46, 121, 152, 163, 183, 186, 188
  device, 3
Monoclinic, 150, 156, 159-160, 163
Morphology, 68, 109, 123, 129, 156
Mould, 183
Moulding, 5, 106, 183, 188
Muco-adhesive, 106
Multiplicative scatter correction (MSC), 32-33

# N

N-vinylpyrrolidone (VP), 151-152, 154-160, 162-163
Nanotechnology, 44
Naproxen, 65
National Science Foundation's Engineering Research Center, 20
Near-infrared (NIR), 4, 9, 21, 30-33, 36, 42, 45, 83, 90, 127-128, 151-153,
    163-164, 183, 188
  spectra, 31, 42, 83, 152, 163-164
  spectroscopy, 9, 31-32, 36
Neusilin®, 60, 63-65, 71
  -US2, 64-65
Newtonian fluid, 191
Nicotinamide (NCT), 76, 79-80, 89-93
Nishi–Wang, 151, 155
Nitrogen, 100, 123, 140, 153
  atmosphere, 123, 153
Noran 7 software, 123
Nuclear magnetic resonance (NMR), 83, 102
Nucleation, 80, 86, 90
Nuclei, 126
Nurofen®, 67, 100
  Meltlets Lemon, 67, 100
NuvaRing, 43, 57

# O

Off-line monitoring, 36
Olympus BX60 microscope, 124, 140, 153
Olympus Corp., 124, 141, 153
On-line monitoring, 8, 28, 44, 121
OpenFoam®, 37
Optical, 131, 143, 149
Optimisation, 8, 11, 20, 26, 29, 35-36, 38, 40, 102, 104, 115, 127, 139, 149,
    181-182
Optimise, 14, 20, 23, 53, 58, 93, 106, 171, 185, 189, 193
Optimised, 9-10, 14, 20-21, 36-37, 41, 53, 64, 91, 102, 115, 126-127, 142, 144,
    146, 184, 188-189
Optimising, 10, 53, 98, 106, 109, 186
Oral, 2, 54-55, 57-58, 60, 62, 97-98, 105-106, 121-122, 129, 139, 175, 181, 187
  bioavailability, 2, 122, 181
  cavity, 106
Orally disintegrating tablet(S) (ODT), 66-67

Organic, 20, 28, 54-55, 64, 75-76, 79, 93, 105
  solvent, 55, 75-76, 79, 93
Orthorhombic, 70-71, 150, 159-160, 162-163
Outer diameter (OD), 189-190

**P**

Paddle, 66, 124, 141, 146
Palatability, 98-99, 104-105
Paracetamol (PMOL), 98, 101-102, 109, 121, 150-164
Parsum, 188
Partial least square(s) (PLS), 31-34, 45, 102, 112, 114
Particle(s), 2, 6-8, 10, 23-25, 36-37, 41, 56, 63, 68, 78, 92, 98, 109, 123-125, 129,
    134, 139, 141, 145, 156, 183, 188
  engineering, 8, 63, 68, 98
  engineering/coating, 98
  matrix, 129
  shape, 23
  size, 2, 7, 10, 23, 36, 41, 78, 124, 139, 141, 145, 156, 183, 188
    distribution, 36, 78, 124, 141, 188
    reduction, 139
Particulate, 20, 55
  synthesis, 20
PDF-2 (X-ray diffraction database), 156, 166
Pellet, 104
Pelletised, 5
  feed, 5
Pelletiser, 22, 58, 123, 128, 176-177, 179
Pelletising, 22, 184, 191
  pH, 4, 46, 59, 64, 78, 100, 111, 113-114, 124, 132-133, 141, 146
Pharma, 7, 12-13, 43, 107, 169, 171, 173, 175, 177, 179
Pharmaceutical, 1-2, 4-10, 14-21, 26-32, 35, 38-40, 43-45, 47-55, 57, 59-61, 63,
    65, 67-73, 75, 77, 79, 81, 83, 85, 87, 89, 91-95, 97-99, 101, 105-107, 109,
    111, 113, 115-119, 121, 123, 125, 127, 129, 131, 133, 135-137, 139, 141,
    143, 145, 147, 149-151, 165-167, 169-171, 174-182, 184-186, 188-189,
    191-195
  analysis, 107
  dosage form, 2, 125
  engineering industry, 31
  environment, 169, 171
  industry(ies), 1-2, 5, 7-9, 14, 19-20, 26, 28, 35, 50, 53-54, 59, 68, 98, 121, 149,
    169, 174-175, 178, 181-182, 185-186, 192
  manufacturing process, 39, 70, 192

Pharmaceutics, 2-3, 15-18, 47-52, 59-62, 71-73, 86-87, 93-95, 101, 115-118, 136-137, 147-148, 165-166, 179, 193-194

Pharmacokinetics, 68

Pharmacology, 49, 51-52, 59-60, 94, 147

Phase diagram, 75, 88

Phase distribution, 123

Phase separation, 69, 150

Phase transformation, 71

Phenacetin, 65

Photometric, 12

Physical, 6, 8, 24, 28-29, 37, 41, 59, 63-64, 66, 70, 83-85, 92, 98, 105, 110-111, 123, 130, 139-140, 149-151, 157, 187, 191
  mixture(s) (PM), 41, 46, 59, 66, 70, 83-85, 91, 92, 105, 110-111, 123, 130-132, 140-141, 144, 151, 153, 159-163
  properties, 28, 149

Physicochemical, 8, 36, 68, 70, 75-76, 102, 110, 156, 184
  properties, 68, 75, 110

Physics, 37, 61

Physiological, 69, 75

Pigment, 12

Pipe(s), 5, 32, 54

Pipeline, 8, 63, 68

Piroxicam (PRX), 106

Placebo, 101-102, 107, 112-114

Plasdone® S630, 68, 151

Plastic(s), 5, 23, 54
  industry, 54

Plasticisation, 68

Plasticised, 23, 54

Plasticiser, 6, 54, 99-100, 102, 104-105, 111, 142

Plasticising, 26, 54, 67, 100

Polar, 109

Polarisation, 86, 151

Poly(4-methyl 1-pentene), 85

Poly(ethylene-*co*-vinyl acetate) (EVAC), 6-7

Polyelectrolyte complex, 98

Polyether ether ketone, 175

Polyether ketone ketone, 175

Polyethylene, 6-7, 26, 57, 65, 104, 171
  glycol(s) (PEG), 6, 26, 43, 65, 104
  -8000, 65
  oxide(s) (PEO), 7, 171

Polyglycolide, 6
Polylactide, 6
Polymer, 5-8, 16, 18, 20, 24, 26, 28, 33, 36-38, 42, 47-48, 54-59, 61, 65-68, 71,
    92-93, 97-102, 105-106, 108-112, 114-115, 122, 126-133, 135, 139-140,
    142-144, 146, 150-152, 155-159, 162, 166, 171-173, 185, 191, 193-194
  carrier, 6, 92-93
  chain, 126
  compound(ing), 127, 171
  matrix, 5, 7, 26, 98, 100, 110, 114-115, 131, 150, 157-159
Polymeric, 6, 19, 21, 26, 63, 68, 97-98, 100, 102, 104-106, 111, 118, 122, 124-
    126, 128-129, 132-133, 139-140, 150, 153, 169, 183, 191
  carrier, 124, 140, 153, 183
  coating, 132-133
  matrix, 102, 104, 125-126, 128, 150
Polymerisation, 152
Polymorph, 70, 149, 159
Polymorphic, 36, 69-71, 149-151, 153, 155-157, 159-163, 165, 167
  transformation, 36, 70-71, 150, 156, 160, 162-163
Polymorphism, 70, 149
Polystyrene, 5
Polyvinyl acetate, 26
Polyvinyl caprolactam, 68
Polyvinyl pyrrolidone (PVP), 6, 101
Poorly soluble drug, 58
Poorly water-soluble drug, 67, 70, 139, 146
Position sensitive detector (PSD), 124, 141, 153
Powder, 21-22, 35, 39, 41, 44, 51, 58, 66, 70, 87, 91, 100, 105, 124, 128, 140-
    141, 144-145, 150, 152, 156, 170, 183, 187-188, 191
  ppm, 86, 88
Praziquantel, 104-105
Precipitation, 75, 150
Precirol® ATO 5, 105
Premix(ing), 64, 191
  feeding, 191
Preparation, 1, 108, 121-122, 135, 149-150, 154
Preparat® 4135F, 100
Press, 1, 16, 41, 187
Pressure, 1, 3-4, 12, 21-23, 25-26, 31, 46, 88, 106, 149, 171-173
  flow, 23
Price, 19, 28, 186
Principal component analysis (PCA), 32-34, 45, 101, 112, 123
Probe, 32-33, 83, 128, 152-153, 183, 188

Process analytical technology (PAT), 1, 9, 34, 38, 45-46, 128, 170, 186, 188, 192
Process parameter, 14, 18, 46
Processing conditions, 6, 98, 127-128, 184, 193
Processing temperature, 7, 76, 80, 83, 85-87, 89-90, 92-93
Production, 1, 4-5, 8-9, 11-14, 19, 21, 24, 27, 29-32, 34, 37-38, 40, 44, 54, 76,
    97, 129, 131, 149-150, 171, 175-176, 181-182, 185, 187-188
  cost, 9
Progesterone, 65
Proportional–Integral–Derivative (PID), 39-40, 42, 151
Propranolol hydrochloric acid (PRP), 100, 102
Protonated, 100, 127
Pulverisette ball milling system, 152
Pumping, 24-25
Purification, 76, 78
Purity, 29, 75, 78, 89, 157

**Q**

Quality, 1, 5, 8-10, 17-20, 23, 26-27, 30-31, 35-38, 40-41, 44-47, 52, 64, 72-73,
    121, 128, 170, 172, 182, 184-189, 191-195
  assurance, 8, 27, 46-47, 52, 121, 182, 184, 192, 194-195
  -by-design (QbD), 9-10, 35-36, 186, 192
  control, 1
  management, 44
Quantum mechanical (QM), 10-11, 126, 131
Quaternary, 104

**R**

Raman, 4, 30-33, 36, 42, 45, 50, 56, 91, 183, 188
  mapping, 56
  spectroscopy, 36, 50, 91
Rate constant, 124-125
Recrystallisation, 20, 105
Regulation, 30
Regulatory compliance, 27-28
Reheating phase, 171
Relative humidity (RH), 79
Release, 1-2, 4-7, 15-16, 28, 41, 44, 46, 53-55, 57-59, 66-68, 70, 93, 97, 101,
    104-106, 114, 117-118, 121-122, 124-126, 132-137, 141, 146, 166, 170,
    175, 180
  constant, 125
  rate, 2, 57-58, 122, 125, 134-135, 175
Reproducibility, 75, 99

Repulsion energy, 126
Research, 2-3, 7-8, 15, 17-18, 20, 27, 41, 47-49, 59-60, 63, 70-72, 79, 93-95, 101, 115-116, 119, 121, 147, 149, 165-166, 171, 179, 191, 194
Residence time, 10, 12-13, 21-22, 31, 79-80, 83, 90-91, 172-174, 185
    distribution (RTD), 10-13, 185-186, 189
Retarded-release, 7, 121, 132-133
Retsch GmbH, 58, 123, 152
Roll-mixing, 139
Roller compaction, 19, 39
Rondol Technology Ltd., 171, 174-175, 177
Room temperature, 75, 104-105
Rotate, 25
Rotating, 3, 9, 12, 18, 21-22, 24-26, 37, 41, 53-54, 69, 105, 176, 178, 190
    screw, 3, 21-22, 24, 41, 53, 190
    speed, 26
Rotation, 22, 25, 99, 124, 141
Rotor, 58, 123, 183, 188
    mill, 123, 183, 188
rpm, 3, 12, 42, 58, 76, 78, 80, 83, 90-91, 123-124, 128, 140-141, 146, 152, 173-174, 189
Rubber, 1, 5
Rubbery, 26
    state, 26

## S

Saccharin (SAC), 78-79, 81-85, 89-90
Safety, 8, 44
Salt, 98, 100, 139
    formation, 139
Saltiness, 98
Scale-up methods, 11
Scanning electron microscopy (SEM), 58-59, 65, 68, 77, 109, 123, 129, 156
Scattering, 36
Schering–Plough, 57
Schuler, 13
Screening, 5, 8, 63-64, 121, 131, 139
Screw, 3-4, 9-13, 18, 21-26, 37-38, 41, 53, 58, 64, 69-70, 76, 78-80, 83, 88-91, 93, 99, 105, 115, 122, 127-128, 140, 142, 152, 171-175, 178-179, 184, 188-191
    configuration, 26, 37, 127, 171, 173-175
    design, 24-25, 76, 90, 93, 189
    diameter, 11, 13, 23, 128, 175, 189-190

fill, 189
geometry, 3-4, 9, 23, 37, 69
rotation, 25, 99
size, 24
speed, 3, 9-13, 58, 76, 80, 83, 90-91, 93, 115, 122, 127-128, 142, 152, 171, 173-174, 184, 189
surface, 23
Seal, 177
Sealed, 123, 140, 153
Sensitivity, 99
Sensor, 101, 107
Sensory, 102, 111-112, 114
Shape, 3-4, 7, 21, 23-24, 38, 41, 53-54, 135, 149, 156, 169
Shaped, 43
Shaping, 21, 37, 169-170, 177
zone, 37
Shear, 3, 6, 9, 21, 24-25, 35, 38, 56, 69, 71, 79-80, 83-84, 90-91, 98, 127, 131, 171, 173, 187, 189-190
cell, 71
force, 3, 21, 79-80, 84, 90-91, 127, 189
intensity, 79-80, 83, 90
mixing, 98
Shearing, 22, 24
ShinEtsu Chemical Co., Ltd, 140, 142
Shrinkage, 57
Shrinking, 57
Siemen's PAT (SIPAT), 34, 182, 188
Sigma Aldrich Corp., 107, 140
Silicon, 104, 153
dioxide, 104
Simulation, 36-37, 40-41, 99
Single-screw, 3, 9, 24, 69, 78
extruder, 3, 9, 24, 69
extrusion, 78
Slugging, 19
Slurry conversion, 75
Small-angle X-ray scattering, 36
Sodium benzoate, 104
Sodium orthophosphate, 124, 141
Soft, 7, 34, 107, 175, 177
extrusion, 175, 177
Softening, 5

Solid, 1, 4-8, 19-24, 26, 28, 32-33, 37, 40, 42-44, 53, 57-58, 60, 63, 65-71, 73, 75-76, 79, 83, 92-93, 98, 100-102, 104-106, 109-111, 121, 124, 126, 130-132, 135, 141-142, 144, 149-151, 154, 163, 170, 175, 181, 187, 191
   bed, 24
   dispersion, 1, 7, 20, 37, 42, 58, 65-66, 93, 131-132, 150, 181
   dosage form, 28
   paraffin, 104
   -phase lipid extrusion, 104-105
   solution, 1, 6, 20, 37, 149
   -state, 28, 32-33, 58, 63, 65, 67, 69, 71, 73, 75-76, 79, 83, 100-102, 106, 109-110, 124, 130, 141-142, 149-150
   analysis, 101, 109
   method, 75
   nuclear magnetic resonance (SSNMR), 83, 88
SOLIQS, 7
Soller slit, 153
Soluble, 1, 4-5, 8, 20, 55, 58, 63, 67-68, 70, 75, 77, 79, 81, 83, 85, 87, 89, 91, 93, 95, 100, 105, 139, 146, 163, 181
   drug, 58, 67, 70, 139, 146
Solubilisation, 92, 131, 139, 143, 159
Solubilise, 26
Solubility, 1, 5, 8, 10, 20, 28, 36, 42, 54, 58, 63-66, 68-70, 75, 79, 91-92, 108-109, 126, 131, 139, 142, 149, 151-152, 154-155, 159, 166, 170
   parameter, 36, 108-109, 126, 151, 154-155
Soluplus® (SOL), 11-12, 68, 91-92, 151-152, 154-164
Solution, 1, 6, 20, 37, 40, 67, 78, 100, 107, 112, 124, 141, 149, 182, 190-191
Solvent, 1, 9, 14, 42, 54-55, 69-70, 75-79, 85-87, 91, 93, 104-106, 139, 150, 170, 181
   -based method, 75
   casting, 105
   evaporation, 1, 42, 54, 75
   -free cold extrusion, 104
   growth method, 69
   method, 75-78, 91
Specific energy (SE), 11
Specific mechanical energy consumption (SMEC), 11-13, 185
Specific surface, 104
Spectra, 31, 42, 59, 83, 86, 102, 152-153, 163-164
Spectrometer, 152
Spectrometry, 152
Spectroscopy, 9, 30-32, 36, 47, 49-50, 59, 83, 91, 99, 151, 163

Speed, 1, 3, 9-13, 22, 26, 58, 66, 76, 80, 83, 90-91, 93, 115, 122-123, 127-128, 142, 146, 152, 171, 173-175, 184, 189-190
Spheronised, 37
Spinning, 86
Split-fed system, 191
Split-feeding, 191
  system, 191
Spot Advance Software, 124, 141, 153
Spray coating, 105
Spray drying, 1, 19, 54-55, 69, 98, 139, 150
Stabilise, 6, 64
Stabilisation, 21
Stabilised, 81
Stability, 4, 6-8, 63-64, 68-69, 97, 105, 149-150
Standard deviation (SD), 31, 114
Standard normal variate (SNV), 32-33
Statistical process control, 31, 34
Steam generation, 34
Stearic acid, 104
Steroid hormone (SH), 122, 124, 126, 128-135
Stiffness, 158
Stirring, 108
Storage, 5, 7, 65, 79, 105
Strength, 33, 77, 99, 126-127, 155
Stress, 71, 79
Structure, 40, 54, 57, 59, 71, 102, 156-157
Sugar, 8, 68, 106
Supercritical fluid(s), 69, 98, 139, 150
  processing, 139, 150
  technique, 69
Supersaturation, 91, 105
Supervisory control and data acquisition (SCADA), 182, 188
Suppositories, 1
Supramolecular, 69
Surface, 11, 15, 20, 22-23, 36, 56, 59, 64-66, 77-78, 80, 90, 99, 104, 109, 123, 125, 129, 132, 134, 149, 156, 187
  area, 11, 36, 77-78, 104, 123, 187
  erosion, 22
  free energy, 56
  morphology, 109, 129, 156
  properties, 149

Surfactant, 42
Sustained-release (SR), 1-2, 6-7, 121-122, 127, 135
Sustaining drug release, 28, 54
Swell, 55
Swelling, 20, 24, 55, 126
Synergistic, 139, 146
Synthesis, 20, 44, 79, 82

**T**

Tablet, 7, 20-21, 33, 35, 37, 39-41, 66, 132-135, 187-188
    coating, 35, 188
    press, 41, 187
Tableting, 44, 187
Talc, 67, 100
Targeted-release, 4, 121, 170
Taste, 5, 53, 67, 97-102, 104-109, 111-115, 117, 119, 170
    efficiency, 105
    masked, 97, 99, 104-106, 115
    masking, 67, 97-102, 104-109, 111-115, 117, 119
    perception, 98, 104
    profile, 107
Temperature, 1, 3, 6-7, 9-10, 12-13, 20, 22-23, 26, 31, 37-38, 46, 53, 55-56,
        58-59, 66, 70-71, 75-76, 79-81, 83-90, 92-93, 100, 104-105, 122-124, 127,
        130, 142-143, 149-153, 157-162, 171, 173-175, 179, 188
    distribution, 37
    profile, 9, 55, 58, 152, 173
    program, 12
    pulse, 153
    range, 123, 130, 153, 159
Ternary, 65, 75, 104
Tertiary, 100
Testosterone, 65
Theophylline anhydrous (TA), 86-87
Theophylline-citric acid (TC), 86-87, 89
Therapeutica, 176
Thermal, 6, 28, 59, 66-67, 71, 80, 92-93, 123, 130-131, 142-143, 151, 153,
        158-160, 165, 167, 171
    analysis, 67, 71, 123, 130, 142, 151, 153
    behaviour, 28
    degradation, 92-93, 171
    enthalpy relaxation, 158
    motion, 80

stability, 6
transition, 130, 142-143
Thermo Fisher Scientific Inc., 12, 14, 153
Thermo Fisher, 10, 12, 14, 58, 122-123, 128, 152-153
Thermo Scientific, 18, 152
Thermo-chemical, 6
stability, 6
Thermo-Noran™, 123
Thermodynamic, 20, 56, 79, 149
adhesion, 56
Thermodynamically, 70
Thermogram, 110, 157
Thermoplastic, 6, 37, 57, 121
Time, 2, 4, 7-10, 12-13, 19, 21-22, 27-29, 31, 34-35, 39, 42, 45, 56, 63, 66-69, 79-80, 82-84, 86, 90-92, 97, 99, 104, 121, 123-126, 132-135, 141, 146, 152-153, 158, 170, 172-174, 184-189, 192
-temperature superposition, 80
Torque, 3, 11-13, 21, 26, 31, 127, 189
Toxic, 6, 69
Toxicity, 69
Trans-mucosal, 106
Transdermal, 1, 26, 54
Transformation, 36, 70-71, 149-151, 156, 159-160, 162-163
process, 71
temperature, 160, 162
Transition, 6, 23, 26, 59, 66, 100, 122, 130-131, 142-143, 150, 152, 157-159, 171
rate, 122
time, 158
Transportation, 29, 38, 178
Transverse flow, 23
Triclinic, 70-71
Turbula® TF2 mixer, 58, 122, 140, 152
Twin-screw, 3, 9-10, 12, 18, 24-25, 58, 64, 69-70, 76, 79, 89, 91, 122, 140, 152, 171-175, 178-179, 189-191
extruder, 3, 10, 12, 25, 58, 69-70, 76, 122, 140, 152, 171-174, 189-191
extrusion (TSE), 76-80, 83-93, 171, 175, 179, 189-190
Two-dimensional, 85, 109

**U**

UltraDry, 123
Ultrasonication, 69

Ultrasound, 4
Ultraviolet, 91
University in Belfast, 174
University of Greenwich, 60, 107, 111
University of Istanbul, 174
University of Newcastle, 174
US Food and Drug Administration (FDA), 1, 8-9, 17-18, 27, 30, 47, 49, 52, 72-73,
    121, 170, 175, 179, 181, 186, 191-192, 194-195

## V

Vacuum, 21, 123
van der Waals, 56, 69, 159
van Krevelen, 108-109, 116, 126, 151, 154
  van Krevelen–Hoftyzer, 108, 126, 151, 154
Vapour, 71, 79, 88
  phase, 71
Variable temperature X-ray powder diffraction (VTXRPD), 150-151, 153,
    159-163
Varian 705 DS dissolution paddle, 124, 141
Varian Inc., 124, 141
Vegetable calcium stearate, 7, 121
Venting, 2, 22
Verapamil formulation, 7
Verapamil hydrochloric acid (VRP), 99-102, 106-115
Vertical extruder, 176-178
Vertical parallel, 179
Vertical processing, 177-178
Vessel, 124, 141
Vinyl acetate (VA), 6, 57, 68, 101, 109, 151-152, 154-160, 162-163
Vinyl pyrrolidone, 68
Viscoelastic, 55-56
Viscous, 173
  Viscosity, 6, 13, 21, 26, 36-38, 56, 158, 172-173
Volume, 13, 24, 27, 49, 125, 141, 151-152, 154-155, 157-158, 173, 187
  fraction, 152, 157-158
  -specific feed load (VSFL), 13
  Volumetric scale-up, 10

## W

Water, 4-5, 20, 28, 53-55, 58, 63-64, 67-68, 70, 79, 85-87, 89, 93, 100, 107-108,
    113-114, 133, 139-141, 146, 163, 169, 181
  cooling, 169

-insoluble, 53, 100, 139-140
-insoluble drug, 140
resistant, 63, 68
-solubility, 58
-soluble, 5, 20, 55, 63, 67, 70, 93, 139, 146, 163, 181
drug, 67, 70, 139, 146
Weight fraction, 82
Weight ratio, 91
Wet, 19, 35, 39, 53, 64, 104, 187-188
granulation, 19, 35, 39, 64, 187-188
Wettability, 26, 106
Wide-angle X-ray diffraction (WAXD), 36, 81, 84
Wide-angle X-ray scattering (WAXS), 84-85
Witocan® 42/44 mikrofein, 105
World Medical Association, 107

## X

X-ray analysis, 111, 123, 131, 144
X-ray diffraction, 36, 68, 70, 81, 91, 124, 141, 156, 184
database (PDF-2), 156
X-ray mapping, 123
X-ray photo-electron spectroscopy (XPS), 99, 102
X-ray powder diffraction (XRPD), 58-59, 66, 87, 91, 100, 109, 124, 132, 141, 144, 150, 156, 159-160
XPhase, 123

## Y

Yield, 26-27, 39, 44, 56, 182
Yielding, 85

## Z

Zero order kinetics, 124-125, 133

κ-carrageenan, 104

Lightning Source UK Ltd.
Milton Keynes UK
UKOW05f0137270815

257620UK00001B/9/P

9 781910 242117